Disclaimer

The publisher of this book is by no way associated with the National Institute of Standards and Technology (NIST). The NIST did not publish this book. It was published by 50 page publications under the public domain license.

50 Page Publications.

Book Title: Economics of the U.S. Additive Manufacturing Industry

Book Author: Douglas S. Thomas;

Book Abstract: There is a general concern that the US manufacturing industry has lost competitiveness with other nations. Additive manufacturing may provide an important opportunity for advancing US manufacturing while maintaining and advancing US innovation. Additive manufacturing is a relatively new process where material is joined together layer by layer to make objects from 3D models as opposed to conventional methods where material is removed. The US is currently the primary user of additive manufacturing technology and the primary producer of additive manufacturing systems. Globally, an estimated $642.6 million in revenue was collected for additive manufactured goods with the US accounting for an estimated $468.9 million or 72.9% of global production in 2011. Change agents for the additive manufacturing industry can focus their efforts on three primary areas to advance this technology: cost reduction, accelerating the realization of benefits, and increasing the benefits of additive manufacturing. Significant impact on these areas may be achieved through the reduction in the cost of additive manufacturing system utilization, material costs, and facilitating the production of large products. There is also a need for a standardized model for cost categorization and product quality and reliability testing.

Citation: NIST SP - 1163

Keyword: Additive manufacturing; manufacturing; 3D printing; supply chain; technology diffusion

NIST Special Publication 1163

Economics of the U.S. Additive Manufacturing Industry

Douglas S. Thomas

National Institute of
Standards and Technology
U.S. Department of Commerce

NIST Special Publication 1163

Economics of the U.S. Additive Manufacturing Industry

Douglas S. Thomas
Applied Economics Office
Engineering Laboratory

August 2013

U.S. Department of Commerce
Rebecca Blank, Acting Secretary

National Institute of Standards and Technology
Patrick D. Gallagher, Under Secretary of Commerce for Standards and Technology and Director

Certain commercial entities, equipment, or materials may be identified in this document in order to describe an experimental procedure or concept adequately. Such identification is not intended to imply recommendation or endorsement by the National Institute of Standards and Technology, nor is it intended to imply that the entities, materials, or equipment are necessarily the best available for the purpose.

National Institute of Standards and Technology Special Publication 1163
Natl. Inst. Stand. Technol. Spec. Publ. 1163, 61 pages (August 2013)
CODEN: NSPUE2

Abstract

There is a general concern that the U.S. manufacturing industry has lost competitiveness with other nations. Additive manufacturing may provide an important opportunity for advancing U.S. manufacturing while maintaining and advancing U.S. innovation. Additive manufacturing is a relatively new process where material is joined together layer by layer to make objects from three-dimensional models as opposed to conventional methods where material is removed. The U.S. is currently a major user of additive manufacturing technology and the primary producer of additive manufacturing systems. Globally, an estimated $642.6 million in revenue was collected for additive manufactured goods, with the U.S. accounting for an estimated $246.1 million or 38.3 % of global production in 2011. Change agents for the additive manufacturing industry can focus their efforts on three primary areas to advance this technology: cost reduction, accelerating the realization of benefits, and increasing the benefits of additive manufacturing. Significant impact on these areas may be achieved through reduction in the cost of additive manufacturing system utilization, material costs, and facilitating the production of large products. There is also a need for a standardized model for cost categorization and product quality and reliability testing.

Keywords: Additive manufacturing; manufacturing; 3D printing; supply chain; technology diffusion

Preface

This study was conducted by the Applied Economics Office in the Engineering Laboratory at the National Institute of Standards and Technology. The study provides aggregate manufacturing industry data and industry subsector data to develop a quantitative depiction of the U.S. additive manufacturing industry.

Disclaimer

Certain trade names and company products are mentioned in the text in order to adequately specify the technical procedures and equipment used. In no case does such identification imply recommendation or endorsement by the National Institute of Standards and Technology, nor does it imply that the products are necessarily the best available for the purpose.

Cover Photographs Credits

Microsoft Clip Art Gallery Images used in compliance with Microsoft Corporation's non-commercial use policy.

Acknowledgements

The author wishes to thank all those who contributed so many excellent ideas and suggestions for this report. Special appreciation is extended to Christopher Brown and Simon Frechette of the Engineering Laboratory's Systems Integration Division and to Kevin Jurrens of the Engineering Laboratory's Intelligent Systems Division for their technical guidance, suggestions, and support. Special appreciation is also extended to Dr. David Butry, Dr. Stanley Gilbert, and Dr. Robert Chapman of the Engineering Laboratory's Applied Economics Office for their thorough reviews and many insights and to Ms. Shannon Takach for her assistance in preparing the manuscript for review and publication. The author also wishes to thank Dr. Nicos Martys, Materials and Structural Systems Division, for his review.

Table of Contents

ABSTRACT III
PREFACE IV
ACKNOWLEDGEMENTS V
TABLE OF CONTENTS VI
LIST OF FIGURES VII
LIST OF TABLES VII

1 INTRODUCTION 1
   1.1 BACKGROUND 1
   1.2 PURPOSE 2
   1.3 SCOPE AND APPROACH 2

2 THE U.S. MANUFACTURING INDUSTRY 3
   2.1 THE CURRENT STATE OF THE INDUSTRY 4
   2.2 SCIENCE AND TECHNOLOGY INNOVATION 8
   2.3 ADDITIVE MANUFACTURING 9

3 ADDITIVE MANUFACTURING STAKEHOLDERS 11

4 INDUSTRY USE OF ADDITIVE MANUFACTURING 17
   4.1 PRODUCTS OF ADDITIVE MANUFACTURING 17
   4.2 ADDITIVE MANUFACTURING SYSTEMS 22
   4.3 ADDITIVE MANUFACTURING COSTS 22

5 ADOPTION AND DIFFUSION OF ADDITIVE MANUFACTURING 29
   5.1 THE DIFFUSION PROCESS 29
   5.2 FACTORS OF DIFFUSION 31
   5.3 DIFFUSION OF ADDITIVE MANUFACTURING 32
      5.3.1 Perceived Attributes of Innovation *34*
      5.3.2 Change Agents *36*

6 OPPORTUNITIES FOR CHANGE AGENTS 37

7 CONCLUSION 41

APPENDIX A: SCHEMATIC DATA MAP 47

APPENDIX B: EQUATIONS AND ASSUMPTIONS 51

List of Figures

Figure 2.1: UNSD Manufacturing Value Added, Top Ten Producers 5
Figure 2.2: UNSD Manufacturing Value Added Per Capita, Top Ten Producers............. 6
Figure 2.3: Manufacturing Value Added Compound Annual Growth, 1985-2010 (UNSD) 7
Figure 2.4: Manufacturing Value Added per Capita, Gross Operating Surplus per Expenditure Dollar, and Compensation per Hour, OECD STAN Data............................ 8
Figure 3.1: Manufacturing Supply Chain 12
Figure 4.1: Supply Chain for Additive Manufacturing Products, 2011........................... 19
Figure 4.2: Supply Chain for Additive Manufacturing Systems, 2011 23
Figure 4.3: Cost Distribution of Additive Manufacturing of Metal Parts by varying Factors 25
Figure 5.1: The Logistical S-Curve Model of Diffusion 29
Figure 5.2: Rogers' Model of Adoption (based on probability distribution).................... 30
Figure 5.3: Variables Determining the Rate of Adoption of Innovations........................ 31
Figure 5.4: Forecasts of U.S. Additive Manufacturing Shipments, by Varying Market Saturation Levels 35
Figure 6.1: Impact of Change Agents on the Net Benefits and Return on Investment for Additive Manufactured Products 37
Figure 6.2: Illustration of the Optimal use of Change Agent Funding for Six Alternative Investments 38

List of Tables

Table 3.1: Stakeholders... 13
Table 3.2: Stakeholder Benefits for Adopting Additive Manufacturing 16
Table 4.1: Additive Manufacturing Shipments.. 18
Table 4.2: Pros and Cons in Product Lifecycle Management.. 24
Table 4.3: High Pressure Die Cast Manufacturing Costs vs. Additive Manufacturing Costs (Selective Laser Sintering).. 26
Table 5.1: Forecasts of U.S. Additive Manufacturing Shipments by Varying Market Potential 34
Table A.1: Supply Chain Components 48
Table A.2: Total Supply Chain Values for Industries Relevant to Additive Manufacturing, $million 2011 49
Table A.3: Supply Chain Values for Additive Manufacturing by Industry, $million 2011 50

1 Introduction

1.1 Background

In 2010, the world produced approximately $10.2 trillion in manufacturing value added, according to United Nations Statistics Division (UNSD) data. The U.S. produced approximately 18 % of these goods, making it the second largest manufacturing nation in the world, down from being the largest in 2009. Many products and parts made by the industry are produced by taking pieces of raw material and cutting away sections to create the desired part; however, a relatively new process called additive manufacturing is beginning to take hold where material is aggregated together rather than cut away. Additive manufacturing is the process of joining materials to make objects from three-dimensional (3D) models layer by layer as opposed to subtractive methods that remove material. The terms additive manufacturing and 3D printing tend to be used interchangeably to describe the same approach to fabricating parts. This technology is used to produce models, prototypes, patterns, components, and parts using a variety of materials including plastic, metal, ceramics, glass, and composites. Products with moving parts can be printed such that the pieces are already assembled. Technological advances have even resulted in a 3D-Bio-printer that one day might create body parts on demand.[1,2]

Additive manufacturing is used by multiple industry subsectors, including motor vehicles, aerospace, machinery, electronics, and medical products.[3] This technology dates back to the 1980's with the development of stereolithography, which is a process that solidifies layers of liquid polymer using a laser. The first additive manufacturing system available was the SLA-1 by 3D Systems. Technologies that enabled the advancement of additive manufacturing were the desktop computer and the availability of industrial lasers.

Although additive manufacturing allows the manufacture of increasingly complex parts, the slow print speed of additive manufacturing systems limits their use for mass production. 3D scanning technologies have enabled the replication of real objects without using molds, which can be difficult and expensive. As the costs of additive manufacturing systems decrease, this technology may change the way that consumers interact with producers. The customization of products will require increased data collection from the end user. Additionally, an inexpensive 3D printer allows the end user to produce polymer-based products in their own home or office. Currently, there are a number of systems that are within the budget of the average consumer.

---

[1] Economist. "Printing Body Parts: Making a Bit of Me." <http://www.economist.com/node/15543683>
[2] GizMag. "3D Bio-printer to Create Arteries and Organs." <http://www.gizmag.com/3d-bio-printer/13609/>
[3] Wohlers, Terry. "Wohlers Report 2012: Additive Manufacturing and 3D Printing State of the Industry." Wohlers Associates, Inc. 2012.

1.2 Purpose

Additive manufacturing technology opens up new opportunities for the economy and society. It can facilitate the production of strong light-weight products for the aerospace industry and it allows designs that were not possible with previous manufacturing techniques. It may revolutionize medicine with biomanufacturing. This technology has the potential to increase the well-being of U.S. citizens and improve energy efficiency in ground and air transportation. However, the adoption and diffusion of this new technology is not instantaneous. With any new technology, new standards, knowledge, and infrastructure are required to facilitate its use. Organizations such as the National Institute of Standards and Technology can enable the development of these items; thus, it is important to understand the size and extent of the additive manufacturing industry. Although many organizations provide estimates on the size of the industry, they are often not comparable to widely published industry data and statistics. This report examines the additive manufacturing industry in the U.S. and develops industry data that is comparable to that published by the U.S. Census Bureau. Additionally, it examines the adoption and diffusion of additive manufacturing technologies.

1.3 Scope and Approach

This report focuses on U.S. additive manufacturing; however, there is limited data on the nation's activities in this area. Wohlers[4] estimates that, globally, $1.714 billion in revenue was generated in the primary additive manufacturing market in 2011. This includes $834.0 million for additive manufacturing systems and materials; $642.6 million from the sale of parts produced from additive manufacturing systems; and $236.9 million for maintenance contracts, training, seminars, conferences, expositions, advertising, publications, contract research, and consulting. This report will focus on using these estimates combined with other figures to generate industry data on additive manufacturing that is comparable to industry data published by the U.S. Census Bureau. Data from the Annual Survey of Manufactures and methods developed by Thomas[5] are used in the development of industry data. The report also examines the adoption and diffusion of additive manufacturing by examining costs and unit sales.

There are variations between different types of additive manufacturing processes. These include photopolymer-based systems, powder-based systems, molten material systems, and solid sheet systems.[6] This report does not delve into the economic implications for each system. Rather it approaches additive manufacturing as a whole. Examining these system-related details would require additional research.

---

[4] Wohlers, Terry. "Wohlers Report 2012: Additive Manufacturing and 3D Printing State of the Industry." Wohlers Associates, Inc. 2012.

[5] Thomas, Douglas S. "The Current State and Recent Trends of the U.S. Manufacturing Industry", NIST Special Publication 1142. December 2012. <http://www.nist.gov/manuscript-publication-search.cfm?pub_id=912933>

[6] Gibson, Ian, David Rosen, and Brent Stucker. Additive Manufacturing Technologies. Springer: New York, 2010. 47-50

2 The U.S. Manufacturing Industry

Over time manufacturing processes have changed dramatically. Robotic arms and other machinery have radically changed the manufacturing environment. For instance, just a few decades ago a company such as Standard Motor Products, which produces replacement parts for car engines, had a number of employees who were illiterate. Today, many of the employees at Standard Motor Products not only need to be able to read, they need to know the computer language of the machinery producing the parts.[7, 8] The increase in productivity that is often the result of these changes means fewer employees are needed to make the same products, possibly resulting in lower employment levels in manufacturing. And, while American manufacturing efficiency is improving, other nations have been developing and improving their own manufacturing industries. Emerging economies such as China have gone from producing some manufactured goods to producing a significant amount of goods. Understanding the current state and recent trends of the U.S. manufacturing industry in light of these issues is difficult. Tassey's "Rationales and Mechanisms for Revitalizing U.S. Manufacturing R&D Strategies"[9] and the commentaries that follow it, illustrate that determining the current and future state of U.S. manufacturing is controversial. Some experts have stated that U.S. multinationals have "abandoned" the U.S. and their global expansion "tends to 'hollow out'" U.S. operations while exporting jobs abroad. Others counter that operations and investment of U.S. multinationals are highly concentrated in the U.S. and maintain a large presence while increasing overseas activities.[10, 11, 12]

National economies are often compared to companies competing for market share. This is a common analogy made when discussing the U.S. manufacturing industry; unfortunately, this comparison can be rather misleading.[13, 14, 15, 16, 17] A national economy

---

[7] Davidson, Adam. "The Transformation of American Factory Jobs, In One Company." NPR. January 13, 2012. <http://www.npr.org/blogs/money/2012/01/13/145039131/the-transformation-of-american-factory-jobs-in-one-company?ft=1&f=100>

[8] Davidson, Adam. "Making It in America." *The Atlantic*. January/February (2012). <http://www.theatlantic.com/magazine/archive/2012/01/making-it-in-america/8844/?single_page=true>

[9] Tassey Gregory. "Rationales and Mechanisms for Revitalizing U.S. Manufacturing R&D Strategies." *Journal of Technology Transfer*. 35 (2010): 283-333.

[10] Slaughter, Matthew J. "How U.S. Multinational Companies Strengthen the U.S. Economy." United States Council for International Business. (March 2010). <http://www.uscib.org/docs/foundation_multinationals.pdf>

[11] National Science Foundation. "Asia's Rising Science and Technology Strength." May 2007. <http://www.nsf.gov/statistics/nsf07319/>

[12] Sirkin, Harold L. "Made in the USA Still Means Something." Bloomberg Businessweek. April 10, 2009. <http://www.businessweek.com/managing/content/apr2009/ca20090410_054122.htm>

[13] Krugman, Paul R. "Making Sense of the Competitiveness Debate." Oxford Review of Economic Policy. Vol 12, no. 3 (1996): 17-25. Paul Krugman won the 2008 Nobel Memorial Prize in Economic Sciences for his work on international trade and economic geography.

[14] Krugman, Paul R. "Competitiveness, A Dangerous Obsession." *Foreign Affairs*. Vol 73. Num 2. March/April (1994): 28-44.

[15] The World Economic Forum defines competitiveness of a nation as "the set of institutions, policies, and factors that determine the level of productivity of a country." This definition relates to productivity and is not consistent with the idea of countries competing for market share. World Economic Forum. *The Global*

is the primary supplier of goods and services to its labor force, while a single company, generally, is not the primary supplier of goods and services to its employees. Additionally, a national economy provides the income for the majority of the nation's consumers, while a business, generally, does not provide the income for the majority of its customers. Moreover, a national economy represents a system of exchange in which a company operates as one entity of that system. Companies can go out of business while nations do not. Domestic demand for goods and services constitutes a great proportion of the demand for a nation's domestically-produced products, where the demand for goods and services from a company is primarily external. In addition to these types of analogies, frequently, anecdotal observations are used to characterize the manufacturing industry;[18] however, the insight from these types of observations is somewhat limited, as the manufacturing industry includes hundreds of thousands of establishments with millions of employees making trillions of dollars worth of goods. Anecdotal observations provide a limited narrow scope of the industry that does not necessarily reflect or apply to the industry as a whole.

The primary goal of devoting resources toward manufacturing activities is to receive a form of benefit for oneself and/or for society as a whole. This is true for all industry stakeholders. Investments are often assessed by the resources devoted to the investment and the resources that are yielded from the investment. The return is then compared to the return on other, similar, ventures. Also considered is the extent or size of one's investment. This is the approach that is taken in the following section to examine the manufacturing industry. Specifically, it examines the U.S. manufacturing industry from the stakeholder's return on investment and compares it internationally. This approach provides a systematic examination of the primary goal of devoting resources to manufacturing and sets it in the context of international performance.

2.1 The Current State of the Industry

According to 2010 data from the UN Statistics Division, the U.S. is the second largest manufacturing nation in the world, with China producing just slightly more than the U.S. as seen in Figure 2.1. This figure contains the ten largest manufacturing nations and illustrates the magnitude and significance of the U.S. manufacturing industry to the global and domestic economy. As seen in the pie charts, the U.S. produced 28 % of the world's goods in 1985. This value declined to 18 % in 2010. Although significant, it is important to note that in order for underdeveloped countries to become developed

---

Competitiveness Report. 2010-2011.
<http://www3.weforum.org/docs/WEF_GlobalCompetitivenessReport_2010-11.pdf>
[16] Porter, Michael E. *The Competitive Advantage of Nations*. 1st ed. (New York: The Free Press, 1990).
[17] Porter asserts that competitiveness is measured by productivity and that measuring a country's competitiveness as its share of world markets is "deeply flawed." Porter, Michael E. "Building the Microeconomic Foundations of Prosperity: Findings from the Business Competitiveness Index." In Porter, Michael E., Klaus Schwab, Xavier Sala-i-Martin, and Augusta Lopez-Claros. The Global Competitiveness Report 2003-2004. (New York: Oxford University Press, 2004).
[18] Greenwald, Bruce C.N. and Judd Kahn. Globalization: The Irrational Fear that Someone in China will Take Your Job. (Hoboken, NJ: John Wiley & Sons 2009).

countries, their production and income will need to approach that of the developed world. This, inevitably, results in a decline in the proportion or market share that each developed country represents. In per capita terms, the U.S. is the fifteenth largest producer and far exceeds China (see Figure 2.2). However, the U.S. compound annual growth rate between 1985 and 2010 is 1.1%, putting it well below the 25th percentile of 181 nations as seen in Figure 2.3.

Using input-output analysis, the direct and indirect effects of U.S. manufacturing as a percent of output ranks 38th out of 45 countries; however, it is important to note that this

Figure 2.1: UNSD Manufacturing Value Added, Top Ten Producers

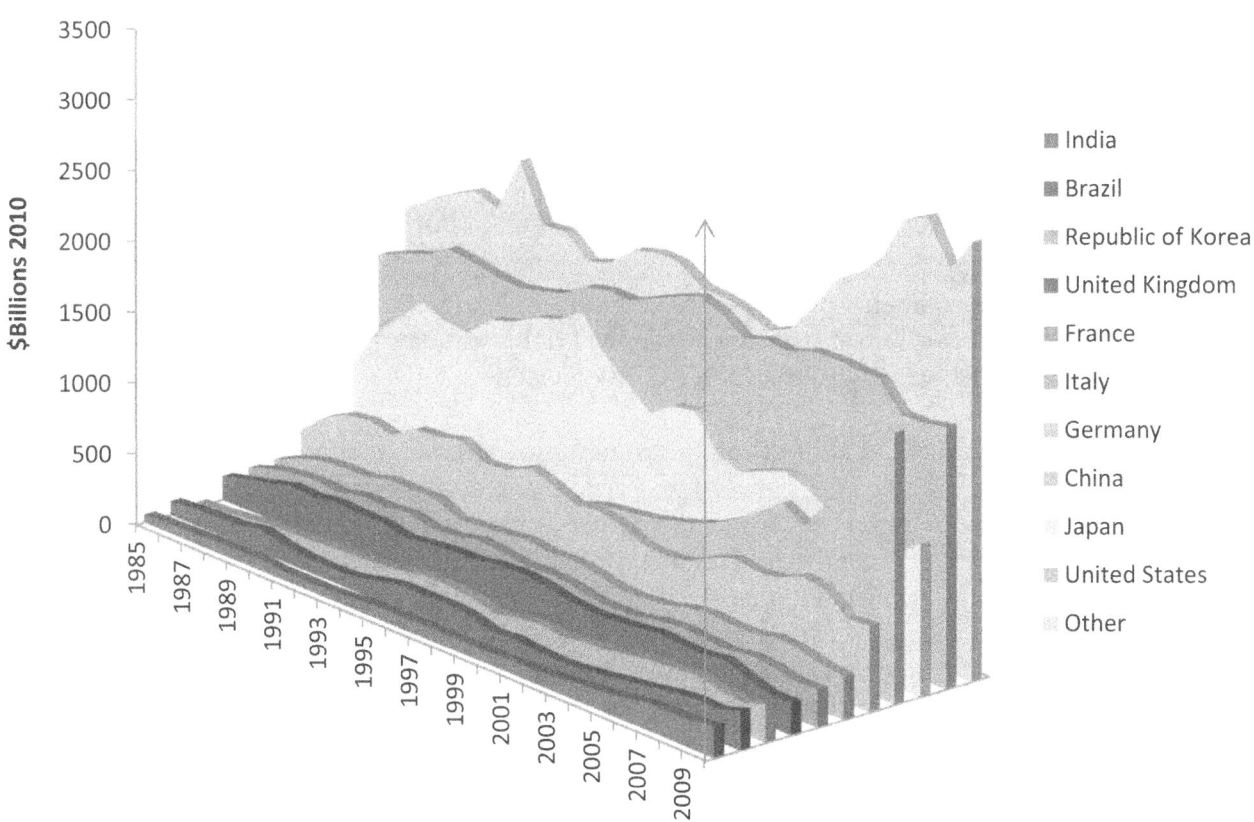

Figure 2.2: UNSD Manufacturing Value Added Per Capita, Top Ten Producers

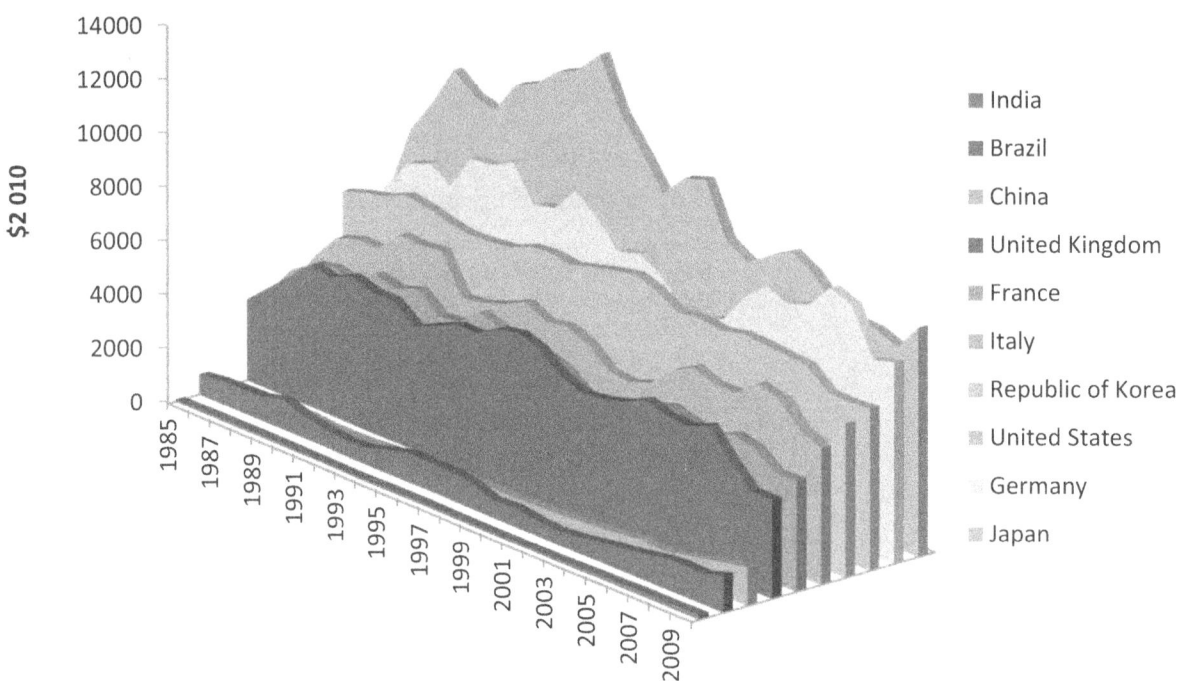

value tends to decrease as nations increase their per capita gross domestic product (GDP). This does not suggest that manufacturing is less important to wealthy nations. While these effects decrease as a percent of output they increase on a per capita basis. Thus, high income nations tend to also have high levels of per capita manufacturing. The correlation coefficient between per capita GDP and manufacturing effects as a share of output is 0.846, suggesting a significant connection.

With the primary goal of devoting resources toward manufacturing activities being to gain a form of benefit for oneself and/or for society as a whole, the best variable to compare the return on investment to owners and financiers is net income per expenditure dollar; however, the primary variable available to examine and compare the returns for owners and financiers internationally is gross operating surplus per dollar of expenditure. Gross operating surplus is gross output less a subset of costs (i.e., intermediate expenditures, compensation, and taxes less subsidies), but does not take into account the depreciation of capital; therefore, it does not fully represent a return on investment. However, it is the best variable available. Employees exchange their time for compensation or income and consumers exchange the purchase price for the utility gained from the product purchased. Unfortunately, data is not readily available to examine the utility of consumers.

Among those countries for which data is available in the Organization for Economic Cooperation and Development's Structural Analysis (OECD STAN) database, Finland and Austria were the only countries to exceed the U.S. in gross operating surplus per expenditure dollar, compensation per hour, and manufacturing valued added per capita (Figure 2.4). Norway, Sweden, Germany, and Denmark have a higher per hour compensation and manufacturing value added per capita than the US, but have a significantly lower gross operating surplus per expenditure dollar. The U.S. manufacturing industry as a whole is just above the 62nd percentile for gross operating surplus per dollar of expenditure, with 14 out of 40 countries having a higher value. Compensation is ranked 9th among 20 countries for which data is available, putting the U.S. at the 55th percentile. The Netherlands, Norway, Sweden, Germany, Denmark, and France have higher levels of per hour compensation. For every dollar of manufacturing value added, there is an estimated 49.5 cents of value added from suppliers of goods and services. The gross operating surplus per expenditure dollar for suppliers was $0.304 for the US, putting it at the 13th percentile. Indonesia had the highest level followed by Turkey, Greece, and Mexico. Compound annual growth in manufacturing between 1985 and 2010 is 1.1% putting it well below the 25th percentile of 181 nations; however, the U.S. continues to be the second largest manufacturing nation in the world, with China producing just slightly more than the US. In per capita terms, the U.S. is the fifteenth largest producer and far exceeds China. Its direct and indirect effects account for 28 % of U.S. output.[19]

Figure 2.3: Manufacturing Value Added Compound Annual Growth, 1985-2010 (UNSD)

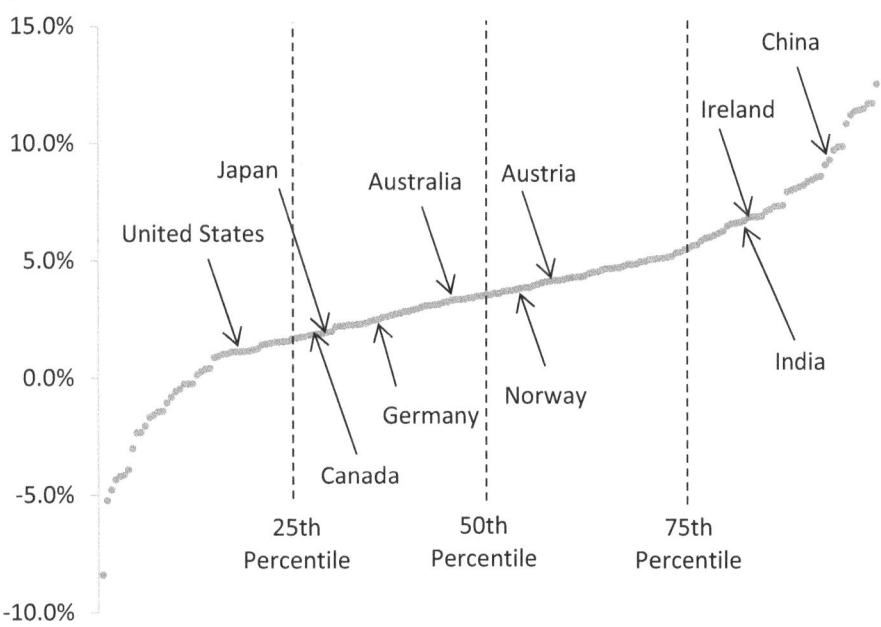

---

[19] Thomas, Douglas. "National Industry Performance Metrics: A Case Study of U.S. Manufacturing." National Institute of Standards and Technology. White paper. Available upon request.

Figure 2.4: Manufacturing Value Added per Capita, Gross Operating Surplus per Expenditure Dollar, and Compensation per Hour, OECD STAN Data

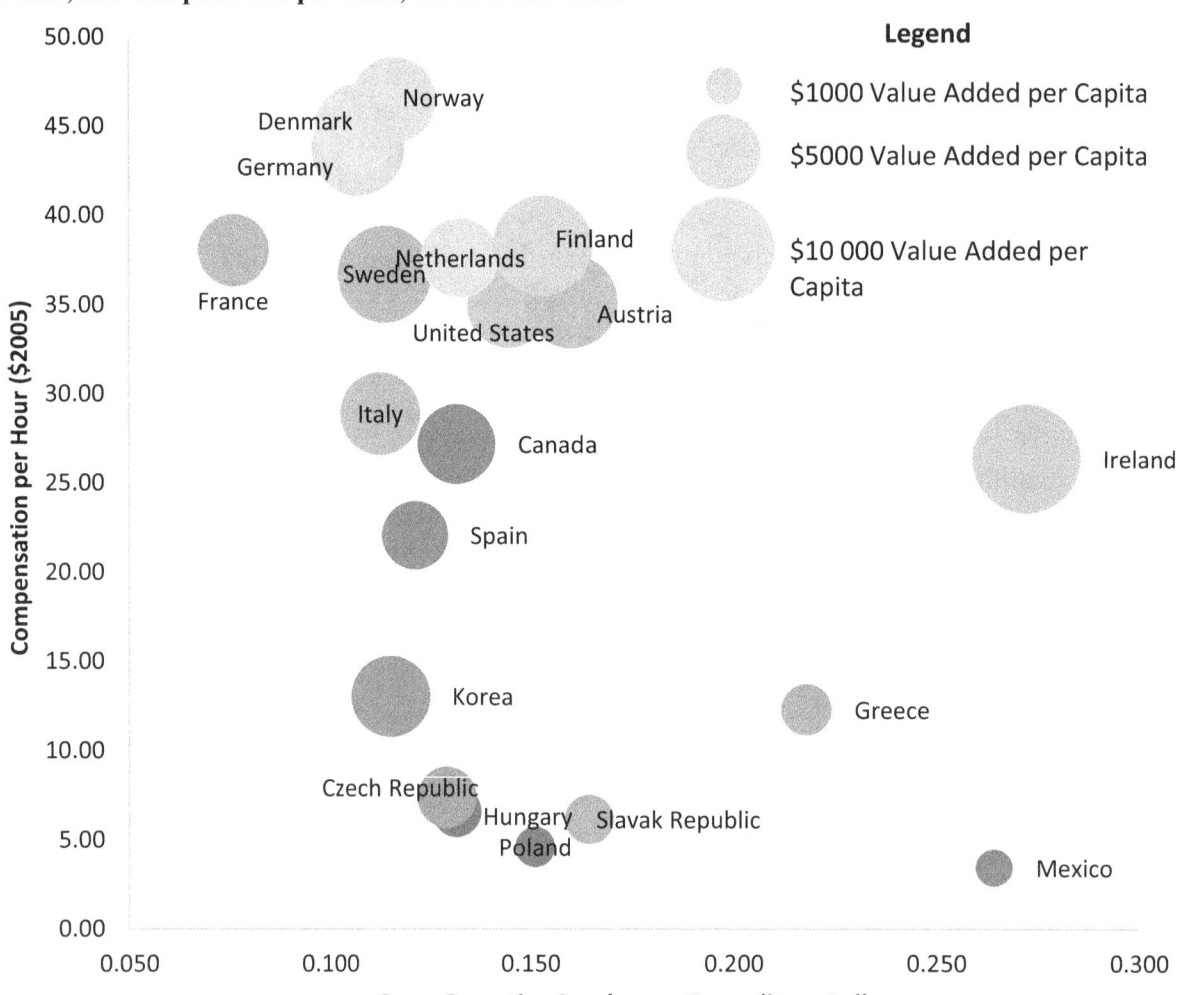

2.2 Science and Technology Innovation

According to the 2012 OECD Science and Technology Outlook, the U.S. is in a lead position for cutting-edge innovation. It maintains excellent higher education and leads the OECD in shares of gross domestic expenditure on research and development (R&D) (41 %), triadic patent families[20] (29 %), and scientific publications (31 %). Large domestic firms contribute to an R&D intensive business sector amounting to 70 % of total gross domestic expenditure on research and development. Small and medium enterprises account for 17 % of business enterprise expenditures on research and development. Approximately 50 % of research and development performers are in high-technology manufacturing.[21]

---

[20] Triadic patent families are defined at the OECD as a set of patents taken at the European Patent Office, Japanese Patent Office, and U.S. Patent and Trademark Office that share one or more priorities.
[21] OECD (2012), OECD Science, Technology and Industry Outlook 2012, OECD Publishing. <http://dx.doi.org/10.1787/sti_outlook-2012-en>

According to adjusted OECD STAN data, the U.S. has the largest research and development expenditure for total manufacturing among those countries for which data is available. In per capita terms, Germany spends nearly as much as the U.S. in research and development for all manufacturing, while Japan exceeds the U.S. expenditure by more than 30 %. Among all OECD countries for which data are available, the U.S. ranks above the 95th percentile for total manufacturing research and development expenditures between 2001 and 2008. From 2001 through 2007, it was above the 90th percentile for all subsectors of manufacturing.[22]

OECD patent data includes the number of patents filed by the inventor's country of residence for 48 countries, including China and India as well as a world estimate. Patents reflect inventive performance and, therefore, are a key measure of innovation. According to OECD patent data, between 1999 and 2007 the U.S. has ranked above the 90th percentile in terms of total number of patents and above the 80th percentile in terms of patents per capita. During that same period, U.S. patents represented between 30 % and 41 % of total patents worldwide. This data is consistent with a patent analysis conducted by Thomson Reuters, which suggested that approximately 40 % of the top 100 global innovator companies are located in the United States.[23] According to the OECD data, Japan is the only country that occasionally produced more patents than the U.S., while Luxembourg, Switzerland, and Japan produced more patents per capita in 2007.[24]

2.3 Additive Manufacturing

There is a general concern that the U.S. manufacturing industry has lost competitiveness with other nations; however, it still maintains a prominent position, as seen in the previous sections. The industry is the second largest in the world, but its growth is below the 25th percentile, placing it under that of Japan, Canada, Germany, and Australia among others. If the current trends in growth continue, by some measures, the U.S. manufacturing industry might be surpassed by other nations. According to the World Economic Forum's Global Competitiveness Index, the U.S. ranked 4th in global competitiveness in 2010-2011, 5th in 2011-2012, and 7th in 2012-2013, setting a downward trend. Another concern is its rank in innovation which in 2012-2013 is 6th, down from 4th in 2010-2011.

Additive manufacturing may provide an important opportunity for advancing U.S. manufacturing while maintaining and advancing U.S. innovation. The U.S. is currently a major user of additive manufacturing technology and the primary producer of additive manufacturing systems. One of the major benefits of this technology is in the area of product design. It allows the production of nearly any complexity of geometry without the need for tooling. Additionally, the complexity does not impact the cost in the same

---

[22] Thomas, Douglas S. "The Current State and Recent Trends of the U.S. Manufacturing Industry", NIST Special Publication 1142. December 2012. <http://www.nist.gov/manuscript-publication-search.cfm?pub_id=912933>
[23] Thomson Reuters. "Top 100 Global Innovators, 2011." <http://www.top100innovators.com/overview>
[24] Ibid

way that it does for conventional manufacturing.[25] This technology eliminates many of the restrictions of 'Design for Manufacture and Assembly,' opening a new realm of possibilities for new customized products at an affordable price point.[26, 27] To some degree, the success of this technology will rely on taking advantage of this benefit. With the U.S. being among the lead innovators and being the primary user of additive manufacturing, this technology may have the potential to significantly impact U.S. competitiveness.

Taking advantage of the opportunities that additive manufacturing offers may prove to be difficult. Designers and manufacturers have established practices and approaches to developing new products. Additive manufacturing presents new possibilities and, to some extent, requires new approaches. Changing the current practices in order to take advantage of new opportunities may be difficult. One such challenge is related to the customization of products to customer needs, which often requires a significant amount of input from the customer. Capturing this information could pose a new challenge to some manufacturers. Although the utility of consumers and end users is difficult to measure, these stakeholders will potentially be a major benefactor of additive manufacturing, as this technology enables rapid design-to-product transformation that enables new products to rapidly come to market.

Unfortunately, the available data does not allow an examination of the return on investment for stakeholders in additive manufacturing at this point in time. Section 4 discusses and estimates values for costs and profit; however, these are only reasonable approximations based on a combination of data sources. A comparison of return on investment using this data would not represent the true state of U.S. additive manufacturing.

Additive manufacturing may make the U.S. a more competitive place for manufacturing resulting in more goods being produced in the U.S.; however, it is important to note that productivity is a contributor to the reduction of manufacturing employment.[28] Even if additive manufacturing results in a significant increase in productivity that attracts jobs from overseas, it may not result in a net increase in manufacturing employment; however, it is possible that additive manufacturing may facilitate a net increase in employment through new products or other means.

---

[25] Hopkinson, Neil, "Production Economics of Rapid Manufacture." In Hopkinson, Neil, Richard Hague, and Philip Dickens. *Rapid Manufacturing*. (Hoboken, NJ: John Wiley & Sons, 2006). 147-157.
[26] Boothroyd, Geoffrey, Peter Dewhurst, and Winston Knight. *Product Design for Manufacture and Assembly*. (New York: Marcel Dekker, Inc, 2009).
[27] Mansour, S., Richard Hague. (2003) "Impact of Rapid Manufacturing on Design for Manufacture for Injection Molding." *Proceedings of the Institution of Mechanical Engineers, Part B: Journal of Engineering Manufacture*.
[28] McKinsey&Company. "Manufacturing the Future: The Next Era of Global Growth and Innovation." November 2012.
<http://www.mckinsey.com/insights/mgi/research/productivity_competitiveness_and_growth/the_future_of_manufacturing>

3 Additive Manufacturing Stakeholders

This section identifies stakeholders and costs related to additive manufacturing. These items are relevant to understanding the adoption and diffusion of this technology. Individual manufacturing stakeholders are affected by the industry in different ways. Therefore, it is useful to identify individual stakeholders and classify them into stakeholder groups. This classification can then be used to identify the primary investment each stakeholder has in the manufacturing industry along with their expected return. Stakeholders evaluate benefits and costs of manufacturing industry investments purely from their "stakeholder" point of view; therefore, it is important to identify each stakeholder's investment and expected return. These perspectives can provide some guidance to the adoption of additive manufacturing.

There are a number of stakeholders for the additive manufacturing industry. The most direct and obvious ones are the owners and employees of manufacturing companies; these are the individuals directly responsible for production. As seen in the manufacturing supply chain in Figure 3.1, there are many suppliers of goods and services that also have a stake in the industry; these include resellers, providers of transportation and warehousing, raw material suppliers, suppliers of intermediate goods, and suppliers of professional services. The items in the figure colored in blue represent suppliers of services, computer hardware, software, and other costs. Tan represents refuse removal, intermediate goods, and recycling, while orange represents machinery, structures, and compensation, with red being the repair of the machinery and structures. Green represents the suppliers of materials. These items all feed into the design and production of manufactured goods that are inventoried and/or shipped. The depreciation of capital and net income are also included in the figure, which affect the market value of shipments. In addition to the stakeholders in the figure, there are also public vested interests, the end users, and financial service providers.

As seen in Table 3.1, stakeholders may have a direct investment in manufacturing, such as industry owners and employees, or an indirect investment through supply chains or industry outputs. Each stakeholder is associated with a primary form of investment. For example, employees invest their labor, while owners invest land and capital. Owners often have labor and/or intellectual property invested as well; however, their primary investment is in the form of land and capital as seen in Table 3.1. Each stakeholder has invested these items with the expectation of receiving compensation or a return on investment. Employees, for instance, expect to be compensated for their labor and owners

Figure 3.1: Manufacturing Supply Chain

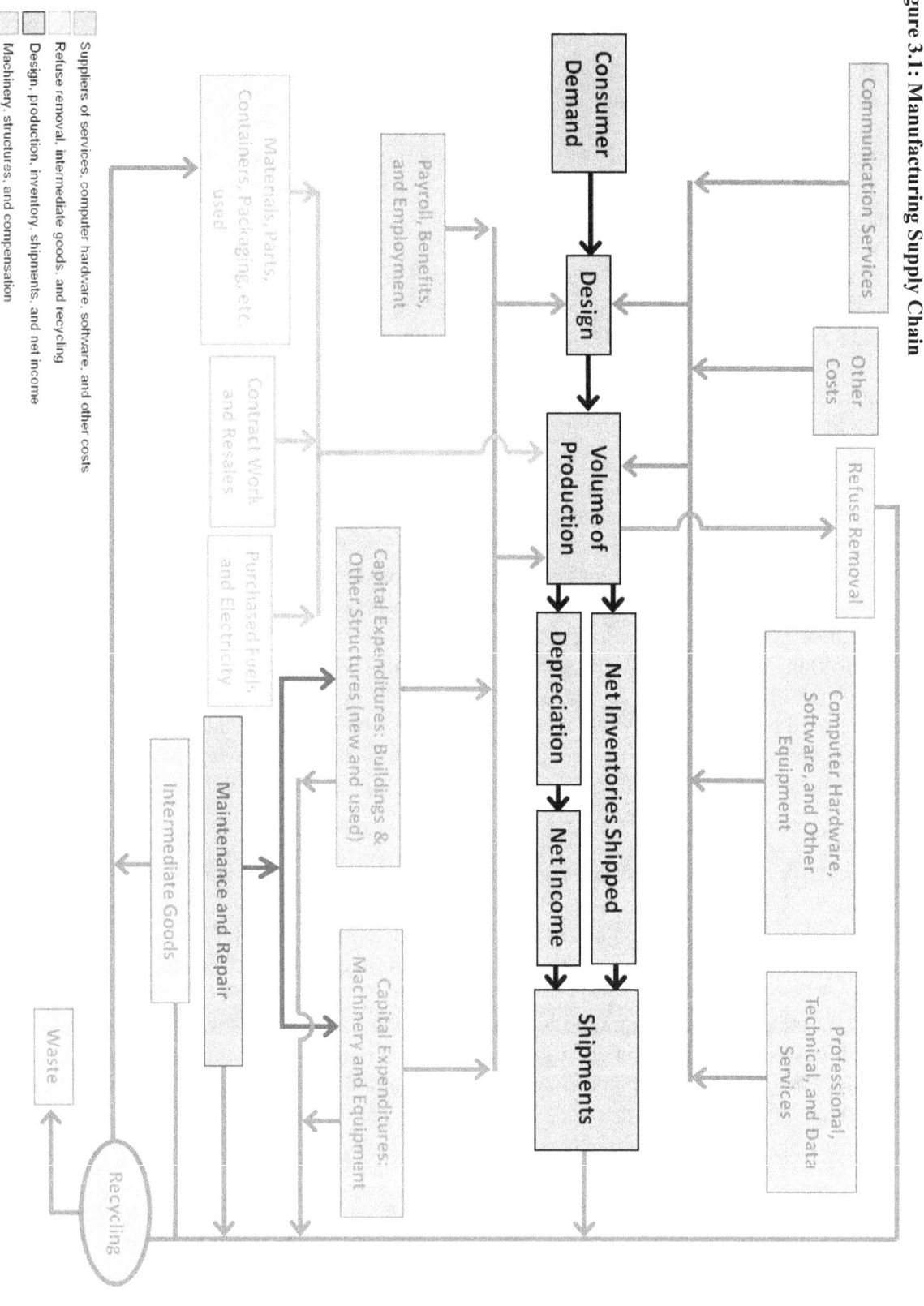

Table 3.1: Stakeholders

Stakeholders	Affiliation	Primary Investment	Expected Return
Owners	Private Producers	Land, Capital Goods, and Financial Capital	Profit From Sales
Employees (manufacturing industry and suppliers)	Laborers	Labor	Income
Resellers			
Retailers	Private Distributer	Land, Capital Goods, and Labor	Profit From Markup
Wholesalers	Private Distributer	Land, Capital Goods, and Labor	Profit From Markup
Standards and Codes Organizations	Public/Private Interest	Labor and Intellectual Property	Economic Success
Transportation and Warehousing	Support Service		
Air Transportation Providers	Transportation	Land, Capital Goods, and Labor	Profit From Fees
Ground Transportation Providers	Transportation	Land, Capital Goods, and Labor	Profit From Fees
Warehousing and Storage Providers	Storage Facility	Land and Capital Goods	Profit From Fees
Professional Societies	Public/Private Support Services	Labor and Intellectual Property	Economic Success and Profit from Fees
Finance Services	Insurance and Finance	Financial Capital	Profit From Fees
Insurance Providers	Insurance	Financial Capital	Profit From Fees
Health and Medical Insurance Providers	Insurance	Financial Capital	Profit From Fees
Financiers	Financier	Financial Capital	Capital Gains
Public Vested Interests			
Policy Makers	Public	Labor and Financial Capital	Economic Success
TaxPayers	Public	Labor and Financial Capital	Economic Success
Industry Suppliers	Public/Private Suppliers	Land, Capital Goods, and Labor	Profit
Mining Material Suppliers	Private Suppliers	Land, Capital Goods, and Labor	Profit From Sales
Agriculture Product Suppliers	Private Suppliers	Land, Capital Goods, and Labor	Profit From Sales
Electric Utility Suppliers	Private Suppliers	Land, Capital Goods, and Labor	Profit From Sales
Water Utility Suppliers	Public/Private Suppliers	Land, Capital Goods, and Labor	Profit From Sales
Natural Gas Suppliers	Private Suppliers	Land, Capital Goods, and Labor	Profit From Sales
Facility Construction Providers	Private Suppliers	Land, Capital Goods, and Labor	Profit From Sales
Maintenance and Repair Providers	Private Suppliers	Land, Capital Goods, and Labor	Profit From Sales
Communication Services Providers	Private Support Services	Land, Capital Goods, and Labor	Profit From Fees
Other Fuel Suppliers	Private Suppliers	Land, Capital Goods, and Labor	Profit From Sales
Refuse Removal Service Providers	Private Support Services	Land, Capital Goods, and Labor	Profit From Fees
Professional Services	Public/Private Support Services	Land, Capital Goods, Labor, and Intellectual Property	Profit From Fees
Legal Service Providers	Public/Private Support Services	Labor	Profit From Fees
Information Service Providers	Public/Private Support Services	Labor	Profit From Fees
Research Organizations	Public/Private Suppliers	Labor and Intellectual Property	Profit From Fees
Accounting Service Providers	Private Support Services	Labor	Profit From Fees
Engineering Service Providers	Private Support Services	Labor and Intellectual Property	Profit From Fees
Computer Service Providers	Private Support Services	Labor	Profit From Fees
Scientific and Technical Service Providers	Private Support Services	Labor and Intellectual Property	Profit From Fees
Advertisers	Private Support Services	Labor and Intellectual Property	Profit From Fees
Other Professional Services	Private Support Services	Labor and Intellectual Property	Profit From Fees
Consumers/End User	End User	Product Purchasing Price	Final Product Utilization

expect to receive a profit. There are six different categories of assets used in Table 3.1 that can be vested into the industry: financial capital, capital goods, land, labor, intellectual property, and the end users purchasing price. A successful industry might be considered one that has a suitable magnitude of production that results in competitive net benefits for its stakeholders. The expected returns from the industry include profits from sales, markup, or fees; income; industry success; capital gains; and utility from the final use of the product.

Summary of Primary Investments

Land: Naturally-occurring goods such as water, air, soil, mineral, and flora used in the production of products (i.e., the totality of goods or services that a company makes available).

Labor: Human effort used in production, which includes technical and marketing expertise.

Capital Goods: Human made goods used in the production of products.

Financial Capital: Funds provided by investors to purchase capital goods for production of products.

Intellectual Property: Ideas, trademarks, copyrights, trade secrets, and patents used to produce products

Purchasing Price: Market value of products sold

Summary of Expected Returns

Profit from sales: The financial benefit realized when revenues exceed costs and taxes for a product.

Capital Gains: An increase in the value of a capital asset

Income: Compensation for an individual's service or labor

Profit from Markup: The difference between the cost of a product and its selling price.

Economic Success: A constant and suitable magnitude of production resulting in competitive benefits (profits, capital gains, income, and product utilization) for an industry's stakeholders.

Profit from Fees: The financial benefit realized when revenues exceed costs and taxes for a service.

Table 3.2 provides a list of stakeholders and the potential impact additive manufacturing might have on them. The adoption of additive manufacturing is likely to have a significant impact on the consumer/end user, as this technology improves new products and facilitates the rapid production of new products. These individuals will be the primary beneficiaries of customized complex products that meet their individualized needs.

Financiers, employers, and suppliers will benefit from the profit of new product sales; however, some of the new products will be replacing previously produced products and the source of revenue might just shift from one product to another. Additionally, any increased profit commanded from these products will be partially reduced through competition as more companies enter the market. The benefit of new customized complex products, however, will continue to benefit end users. It is possible that some of the largest benefits of additive manufacturing will be realized outside of the manufacturing industry.

Table 3.2 also provides a list of costs to stakeholders, as the development and use of additive manufacturing technology has some costs associated with it. The owners invest in the research and development of this technology and also must purchase new machinery to replace traditional manufacturing machinery. Resellers may have to bear the burden of gathering information from customers for customized products. Some of these costs may be passed on to the consumers/end users through the purchase price.

Table 3.2: Stakeholder Benefits for Adopting Additive Manufacturing

Stakeholders	Primary Benefits of the Adoption of Additive Manufacturing	Primary Costs to the Adoption of Additive Manufacturing
Owners	New product sales, increased efficiency and productivity	Cost of research and development, new machinery costs
Employees (manufacturing industry and suppliers)	Reshoring of jobs, increase in income	Labor, possible decrease in employment
Resellers		
Retailers	New product sales	Cost of gathering consumer data for customized products
Wholesalers	New product sales	Cost of gathering consumer data for customized products
Standards and Codes Organizations	Economic success	Cost of research and development
Transportation and Warehousing		
Air Transportation Providers	Increased demand, reduced vehicle weight	Cost of new products
Ground Transportation Providers	Increased demand, reduced vehicle weight	Cost of new products
Warehousing and Storage Providers	Increased demand, reduced vehicle weight	Cost of new products
Professional Societies	Increased demand	Decreased demand
Finance Services	Economic success	Cost of research and development
Insurance Providers	Profit, product reliability and reduced claims, increased demand	Initial investment
Health and Medical Insurance Providers	Product reliability and reduced claims	Minimal cost
Financiers	Increased demand for services	Cost of research and development
Public Vested Interests	Profit from fees and capital gains	Initial investment
Policy Makers	Economic Success, increased standard of living	Cost of research and development
Tax Payers	Economic Success, increased standard of living	Cost of research and development
Industry Suppliers	Economic Success, increased standard of living	Cost of research and development
Mining Material Suppliers	Increased demand	Cost of meeting increased demand
Agriculture Product Suppliers	Increased demand	Cost of meeting increased demand
Electric Utility Suppliers	Increased demand	Cost of meeting increased demand
Water Utility Suppliers	Increased demand	Cost of meeting increased demand
Natural Gas Suppliers	Increased demand	Cost of meeting increased demand
Facility Construction Providers	Increased demand, new construction materials	Cost of meeting increased demand
Maintenance and Repair Providers	Possible increased demand	Cost of meeting increased demand, possible decrease in demand
Communication Services Providers	Increased demand	Cost of meeting increased demand, possible decrease in demand
Other Fuel Suppliers	Increased demand	Cost of meeting increased demand
Refuse Removal Service Providers	Reduced vehicle weight	Cost of meeting increased demand
Professional Services	Increased demand	Cost of meeting increased demand
Legal Service Providers	Increased demand	Cost of meeting increased demand
Information Service Providers	Increased demand	Cost of meeting increased demand
Research Organizations	Increased demand	Cost of meeting increased demand
Accounting Service Providers	Increased demand	Cost of meeting increased demand
Engineering Service Providers	Increased demand	Cost of meeting increased demand
Computer Service Providers	Increased demand	Cost of meeting increased demand
Scientific and Technical Service Providers	Increased demand	Cost of meeting increased demand
Advertisers	Increased demand	Cost of meeting increased demand
Other Professional Services	Increased demand	Cost of meeting increased demand
Consumers/End User	New product utilization, cost reduction, increased efficiency	Increased purchase price

4 Industry use of Additive Manufacturing

Value added is the best measure available for comparing the relative economic importance of manufacturing among various industries, as it avoids the duplication caused from the use of products of some establishments as materials in others. The Annual Survey of Manufactures, one of the datasets used in this report, calculates value added as the value of shipments less the cost of materials, supplies, containers, fuel, purchased electricity, and contract work (i.e., shipments less the suppliers of materials colored green in Figure 4.1). It is adjusted by the addition of value added by merchandising operations plus the net change in finished goods and work-in-process goods. It is important to note that this calculation of value added varies from that of other organizations. The U.S. Bureau of Economic Analysis (BEA), for example, calculates value added as "gross output (sales or receipts and other operating income, plus inventory change) less intermediate inputs (consumption of goods and services purchased from other industries or imported)."[29] The primary difference is that the Annual Survey of Manufacture's calculation of value added includes purchases from other industries such as mining and construction while BEA and other organizations do not include it (i.e., BEA calculates it as shipments less all costs colored blue, tan, orange, red, and green in Figure 4.1). Since this report uses data from the Annual Survey of Manufactures, it will maintain their method of calculating value added.

Although value added is discussed, most of the figures in this report are in terms of shipments, which is analogous to revenue. This value is used because the data collected on additive manufacturing is in terms of revenue; thus, in order to discuss value added, additional assumptions must be made, which introduces additional imprecision.

4.1 Products of Additive Manufacturing

Globally, an estimated $642.6 million in revenue was collected for additive manufactured goods[30] with the U.S. accounting for an estimated $246.1 million or 38.3 % of global production in 2011.[31] As seen in Table 4.1, these products are categorized as being in the following sectors: motor vehicles; aerospace; industrial/business machines; medical/dental; government/military; architectural; and consumer products/electronics, academic institutions, and other. The consensus among well-respected industry experts is that the penetration of the additive manufacturing market is 8 %;[32] however, as seen in

---

[29] Horowitz, Karen J. and Mark A. Planting. *Concepts and Methods of the U.S. Input-Output Accounts.* Bureau of Economic Analysis. 2006.
[30] Wohlers, Terry. "Wohlers Report 2012: Additive Manufacturing and 3D Printing State of the Industry." Wohlers Associates, Inc. 2012: 129.
[31] This value is calculated with the assumption that the U.S. share of additive manufacturing systems sold equates to the share of products produced using additive manufacturing systems. The share of additive manufacturing systems is available in Wohlers, Terry. "Wohlers Report 2012: Additive Manufacturing and 3D Printing State of the Industry." Wohlers Associates, Inc. 2012: 134.
[32] Wohlers, Terry. "Wohlers Report 2012: Additive Manufacturing and 3D Printing State of the Industry." Wohlers Associates, Inc. 2012: 130.

Table 4.1, goods produced using additive manufacturing methods represent between 0.01 % and 0.05 % of their relevant industry subsectors. Thus, additive manufacturing has sufficient room to grow.

Figure 4.1 provides an estimated supply chain for products of additive manufacturing using the methods documented in NIST Special Publication 1142 combined with some additional assumptions.[33] The estimation method used provides rough estimates; thus, some caution should be used. Additional precision would require further data collection. The items in the figure colored in blue represent suppliers of services, computer hardware, software, and other costs. Tan represents refuse removal, intermediate goods, and recycling, while orange represents machinery, structures, and compensation, with red being the repair of the machinery and structures. Green represents the suppliers of materials. These items all feed into the design and production of manufactured goods that are inventoried and/or shipped. The depreciation of capital and net income are also included in the figure, which affect the market value of shipments. The net income per expenditure dollar (i.e., return on investment) is approximately 0.205; however, this may have significant variation. The total number of employees estimated in U.S. additive manufacturing products is estimated at 658. The following sections discuss the various categories of manufacturing that use this technology.

Table 4.1: Additive Manufacturing Shipments

Category	Relevant NAICS Codes	Percent of Total AM Made Products	Shipments of US Made AM Products ($millions, 2011)*	Total Shipments ($millions, 2011)	AM Share of Industry Shipments
Motor vehicles	NAICS 3361, 3362, 3363	19.5%	48.0	445 289.4	0.01%
Aerospace	NAICS 336411, 336412, 336413	12.1%	29.8	157 700.7	0.02%
Industrial/business machines	NAICS 333	10.8%	26.6	365 734.8	0.01%
Medical/dental	NAICS 3391	15.1%	37.2	89 519.5	0.04%
Government/military	NAICS 336414, 336415, 336419, 336992	6.0%	14.8	32 784.4	0.05%
Architectural	NAICS 3323	3.0%	7.4	72 186.9	0.01%
Consumer products/electronics, academic institutions, and other	All other within NAICS 332 through 339	33.6%	82.7	895 709.8	0.01%
TOTAL	NAICS 332 through 339	100.0%	246.1	2 058 925.5	0.01%

* These values are calculated assuming that the percent of total additive manufacturing made products for each industry is the same for the U.S.as it is globally. It is also assumed that the U.S.share of AM systems sold is equal to the share of revenue for AM products
Note: Numbers may not add up to total due to rounding

---

[33] Each supply chain item is calculated for the NAICS codes listed in Table 4.1 and added together by the categories listed in the table using data from the Annual Survey of Manufactures. The values for additive manufacturing are calculated by assuming that the ratio of each supply chain item to the total value of shipments is the same for additive manufacturing. The ratios are then applied to data in the 2012 Wohlers Report. These assumptions have significant implications for precision; however, they are the best estimates available.

Figure 4.1: Supply Chain for Additive Manufacturing Products, 2011

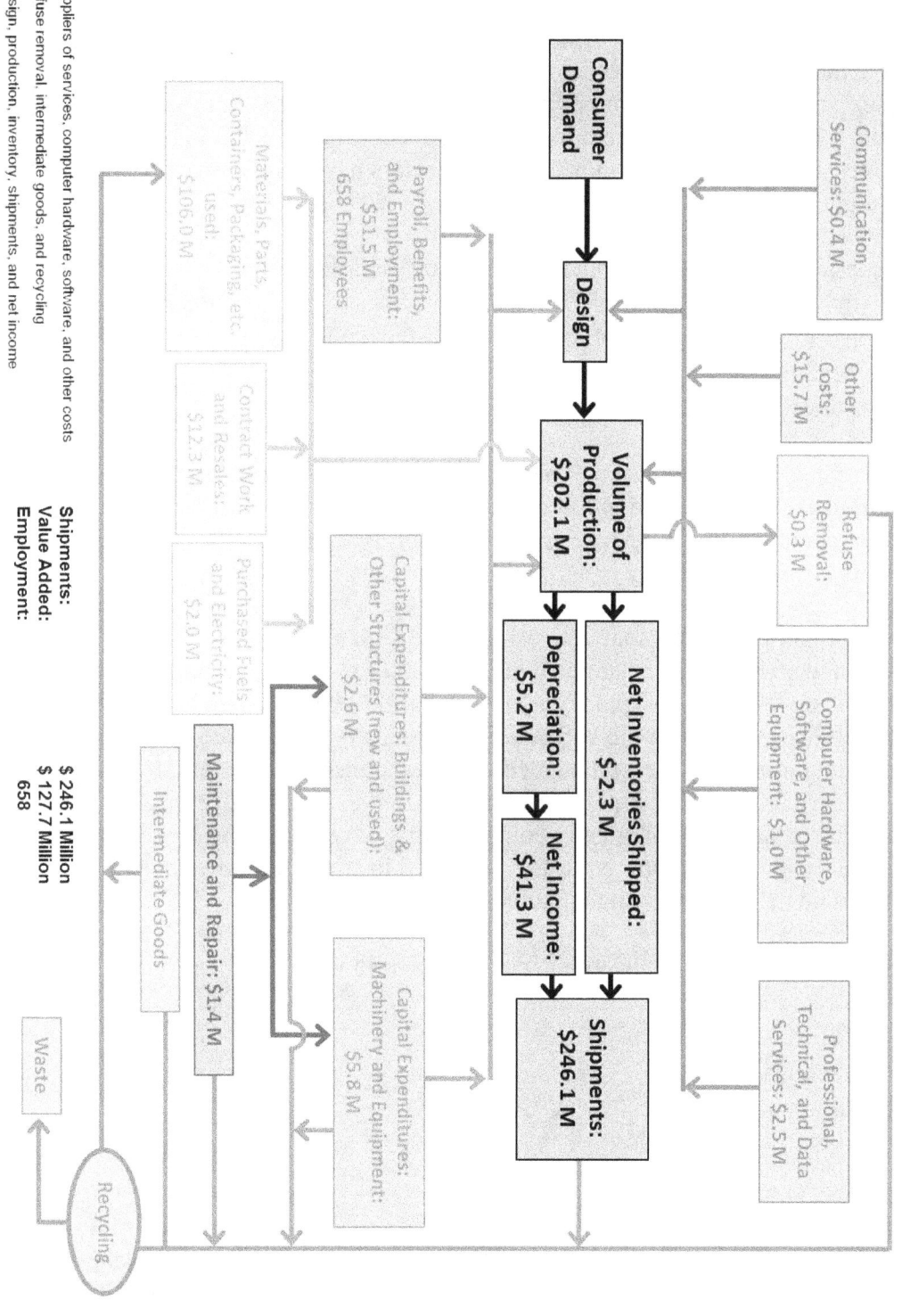

Motor Vehicles: Shipments for the U.S. automotive industry (NAICS 3361, 3362, and 3363) was estimated at $445 billion in 2011. Approximately 19.5% of additive manufacturing is within the automotive industry, with the U.S. share being estimated as $48.0 million or 0.01 % of the U.S. automotive industry. The industry frequently uses additive manufacturing technologies for rapid prototyping. It is also commonly used for complex, high-value, or custom parts for antique cars. Motorsports such as NASCAR and Formula 1 have also been a field for the application of this technology, which have some crossover with the aerospace industry. Both sectors have high demand for performance and weight reduction.

Examples of motor vehicle applications include the following:
- Intake valves, engine bay parts, gear boxes, and engine components
- Air inlet, engine control unit and lower fairing baffle
- Testing of parts
- Motorcycle engines

The restricted construction size of parts made from additive manufacturing has been a limiting factor for further adoption of this technology in the automotive industry. As the additive manufacturing industry develops the ability to produce larger components, the automotive industry is likely to adopt this technology more rapidly.[34, 35, 36]

Aerospace: Shipments for manufacturing in the U.S. aerospace industry (NAICS 336411, 336412, and 336413) were estimated at $157.7 billion in 2011. Approximately 12.1 % of additive manufacturing is within this industry, with the U.S. share being estimated as $29.8 million or 0.02 % of the U.S. aerospace industry. Aerospace includes a range of vehicles including airplanes, unmanned vehicles, transport vehicles, and space vehicles. This industry has significant potential for increased use of additive manufacturing as it often requires strong geometrically complex parts, which must be especially light weight. Additionally, these parts are, typically, produced in small quantities, making them a likely candidate for additive manufacturing.

Examples of aerospace applications include the following:
- Structural parts
- Thrust reverser doors
- Landing gears
- Gimbal eye
- Fuel injection nozzles

---

[34] Gausemeier, Jurgen, Niklas Echterhoff, Martin Kokoschika, and Marina Wall. "Thinking Ahead the Future of Additive Manufacturing – Future Applications." University of Paderborn, Direct Manufacturing Research Center.
[35] Wohlers, Terry. "Wohlers Report 2012: Additive Manufacturing and 3D Printing State of the Industry." Wohlers Associates, Inc. 2012: 130.
[36] Bourell, David L., Ming C. Leu, and David W. Rosen. "Roadmap for Additive Manufacturing: Identifying the Future of Freeform Processing." University of Texas.
<http://wohlersassociates.com/roadmap2009.html>

Similar to the automotive industry, the restricted construction size of additive manufacturing has likely been a limiting factor for further adoption of this technology in the aerospace industry. Additionally, materials, accuracy, surface finish, and certification standards have also played a role in limiting further adoption of this technology.[37, 38, 39, 40]

Industrial/Business Machines: Shipments in U.S. machinery manufacturing (NAICS 333) were estimated at $365.7 billion in 2011. Approximately 10.8 % of additive manufacturing is within this industry, with the U.S. share being estimated at $26.6 million or 0.01 % of U.S. machinery manufacturing. Machinery manufacturing includes the creation of end products that apply mechanical force to perform work. Additive manufacturing technology has been used in the development and production of parts for these machines. For example, a new drag chain link was developed and produced for the mining industry using additive manufacturing.

Medical/Dental: Shipments for U.S. manufacturing of medical and dental products (NAICS 3391) amounted to $89.5 billion in 2011. Approximately 15.1 % of additive manufacturing is within this industry, with the U.S. share being estimated at $37.2 million or 0.04 % of medical/dental manufacturing. The need for custom-made products in the medical and dental industry creates a demand for products made using additive manufacturing methods. Items produced include custom implants, prosthetics, surgical tools, hearing aids, and drug delivery devices among other items. Emerging research and development has resulted in biomanufacturing, where the construction of tissue from living cells is used to "print" organs. Although this field is not fully developed, it is a promising area for applying additive manufacturing technology.

Government/Military: Shipments for U.S. manufacturing of products for the government and military (NAICS 336414, 336415, 336419, 336992) amounted to $32.8 billion in 2011. Approximately 6.0 % of additive manufacturing is within this industry, with the U.S. share being estimated at $14.8 million or 0.05 % of government/military manufacturing. The U.S. military has shown interest in advancing research and procurement of additive manufacturing for a number of components. The U.S. Air Force, for example, is conducting research on the use of additive manufacturing for metal parts, heat exchangers, and plastic resins for remotely piloted vehicles. The U.S. Navy is also investigating the use of this technology.[41]

---

[37] National Institute of Standards and Technology. "Roadmapping Workshop: Measurement Science for Metal-Based Additive Manufacturing." <http://events.energetics.com/nist-additivemfgworkshop/index.html>

[38] Wohlers, Terry. "Wohlers Report 2012: Additive Manufacturing and 3D Printing State of the Industry." Wohlers Associates, Inc. 2012: 130.

[39] Bourell, David L., Ming C. Leu, and David W. Rosen. "Roadmap for Additive Manufacturing: Identifying the Future of Freeform Processing." University of Texas. <http://wohlersassociates.com/roadmap2009.html>

[40] Gausemeier, Jurgen, Niklas Echterhoff, Martin Kokoschika, and Marina Wall. "Thinking Ahead the Future of Additive Manufacturing – Future Applications." University of Paderborn, Direct Manufacturing Research Center.

[41] Scott, Justin, Nayanee Gupta, Christopher Weber, Sherrica Newsome, Terry Wohlers, and Tim Caffrey. "Additive Manufacturing: Status and Opportunities", March 2012. <https://www.ida.org/stpi/occasionalpapers/papers/AM3D_33012_Final.pdf>

Architecture: Shipments for U.S. manufacturing of products for architecture (NAICS 3323) amounted to $72.1 billion in 2011. Approximately 3.0 % of additive manufacturing is within this industry, with the U.S. share being estimated at $7.4 million or 0.01 % of architectural manufacturing. A major use of additive manufacturing for architecture is in the modeling of structures and designs. In the past, physical models were tediously built by hand. Additive manufacturing has revolutionized this process.

Consumer Products/Electronics, Academic Institutions, and Other: Shipments for U.S. manufacturing for consumer products/electronics, academic institutions, and other amounted to $895.7 billion in 2011. Approximately 33.6 % of additive manufacturing is within this industry, with the U.S. share being estimated at $82.7 million or 0.01 % of this category of manufacturing. It includes many items produced using additive manufacturing technology, including toys, figurines, furniture, office accessories, musical instruments, art, jewelry, museum displays, and fashion products among other items.

4.2 Additive Manufacturing Systems

Approximately 62.8% of all commercial/industrial units sold in 2011 were made by the top three producers of additive manufacturing systems: Stratasys, Z Corporation[42], and 3D Systems based out of the U.S. Approximately 64.4% of all systems were made by companies based in the U.S. The total global revenue from system sales was $502.5 million with U.S. revenue estimated at $323.6 million as seen in Figure 4.2.[43] The production of additive manufacturing systems or 3D printers can be categorized as being under NAICS 332: Industrial Machinery Manufacturing. Data from the Annual Survey of Manufactures for this sector was used to develop the estimates in Figure 4.2. The net income as a share of revenue (i.e., shipments) for Stratasys and 3D Systems, two of the three largest additive manufacturing system producers, was 0.144 and 0.178, while the estimate using data in Figure 4.2 is 0.152.

It is important to remember that additive manufacturing systems are already incorporated into the sales of products produced using this technology; thus, it would be unorthodox to add the value for additive manufactured products together with the value for the systems.

4.3 Additive Manufacturing Costs

Manufacturing processes and manufacturing parts are becoming more and more complex. Additive manufacturing both reduces and adds to the complexity of this process. As seen in Table 4.2, there are a number of pros and cons involved with additive manufacturing. For instance, there are fewer parts to manage, more flexibility in design, and products can be individualized; however, there are higher calibration requirements, needed quality

---

[42] Z Corporation was acquired by 3D Systems Inc. in 2012.
[43] The dollar estimate is assumes that the share of U.S. revenue is equal to the share of U.S. unit sales, which is from Wohlers, Terry. "Wohlers Report 2012: Additive Manufacturing and 3D Printing State of the Industry." Wohlers Associates, Inc. 2012: 134.

Figure 4.2: Supply Chain for Additive Manufacturing Systems, 2011

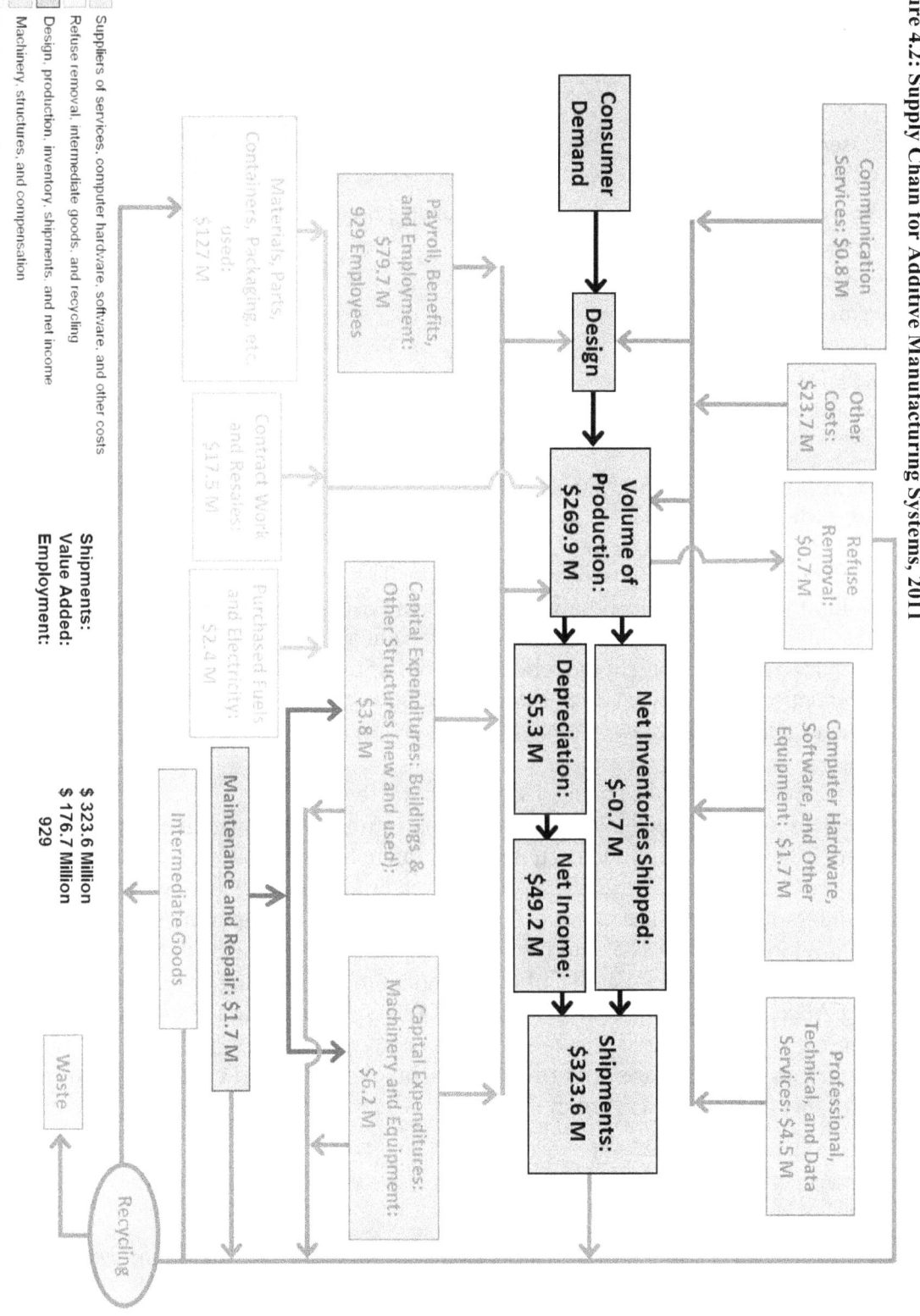

Table 4.2: Pros and Cons in Product Lifecycle Management

Pros	Cons
More flexible development	Software limitations
Freedom of design and construction	High machine and material costs
Integration of functions	High calibration effort
Less assembly	Deficient quality
Fewer production tools necessary	Parts often require reworking
Less spare parts in stock	Building time depends on part height
Less complexity (fewer parts to manage)	
Fewer tools needed	
Less time-to-market	
Rapid alterations	
Individualized products	

Source: Lindemann C., U. Jahnke, M. Moi, and R. Koch. "Analyzing Product Lifecycle Costs for a Better Understanding of Cost Drivers in Additive Manufacturing." Proceedings of the 2012 Solid Freeform Fabrication Symposium. <http://utwired.engr.utexas.edu/lff/symposium/proceedingsArchive/pubs/Manuscripts/2012/2012-12-Lindemann.pdf>

improvements, and parts often require reworking. The benefits of additive manufacturing are not limited to the producer, however, as the end user also benefits from increased functionality, reduced lifecycle costs, and new product utilization. Aerospace parts, for instance, have shown a weight reduction potential of up to 70% of the original part[44] and a 1 kg reduction in weight saves an estimated $3000 of fuel annually[45], not to mention the reduction in emissions.

Costs have been identified as being a significant factor in whether producers adopt additive manufacturing technologies. Hopkinson estimates that machine costs range between 50 % and 75 % of total cost, materials range between 20 % and 40 %, and labor ranges between 5 % and 30 %.[46] The price for materials can vary somewhat. Stereolithography/epoxy-based resin is estimated at $175 per kilogram, selective laser sintering/nylon powder is $75, and fused deposition modeling/ABS filament is around $250. To put this in the perspective of conventional manufacturing, injection molding/ABS is about $1.80 and machining/1112 screw-machine steel is about $0.66.[47]

Other research on metal parts confirms that machine and material costs are a major cost driver for this technology as seen in Figure 4.3, which presents data for a sample part

---

[44] Lindemann C., U. Jahnke, M. Moi, and R. Koch. "Analyzing Product Lifecycle Costs for a Better Understanding of Cost Drivers in Additive Manufacturing." Proceedings of the 2012 Solid Freeform Fabrication Symposium. <http://utwired.engr.utexas.edu/lff/symposium/proceedingsArchive/pubs/Manuscripts/2012/2012-12-Lindemann.pdf>

[45] West, Karl. "Melted Metal Cuts Plane's Fuel Bill." The Sunday Times. Sunday 13 February 2011. <http://www.thesundaytimes.co.uk/sto/business/energy_and_environment/article547163.ece>

[46] Hopkinson, Neil, "Production Economics of Rapid Manufacture." In Hopkinson, Neil, Richard Hague, and Philip Dickens. *Rapid Manufacturing*. (Hoboken, NJ: John Wiley & Sons, 2006). 147-157.

[47] Ibid

Figure 4.3: Cost Distribution of Additive Manufacturing of Metal Parts by varying Factors

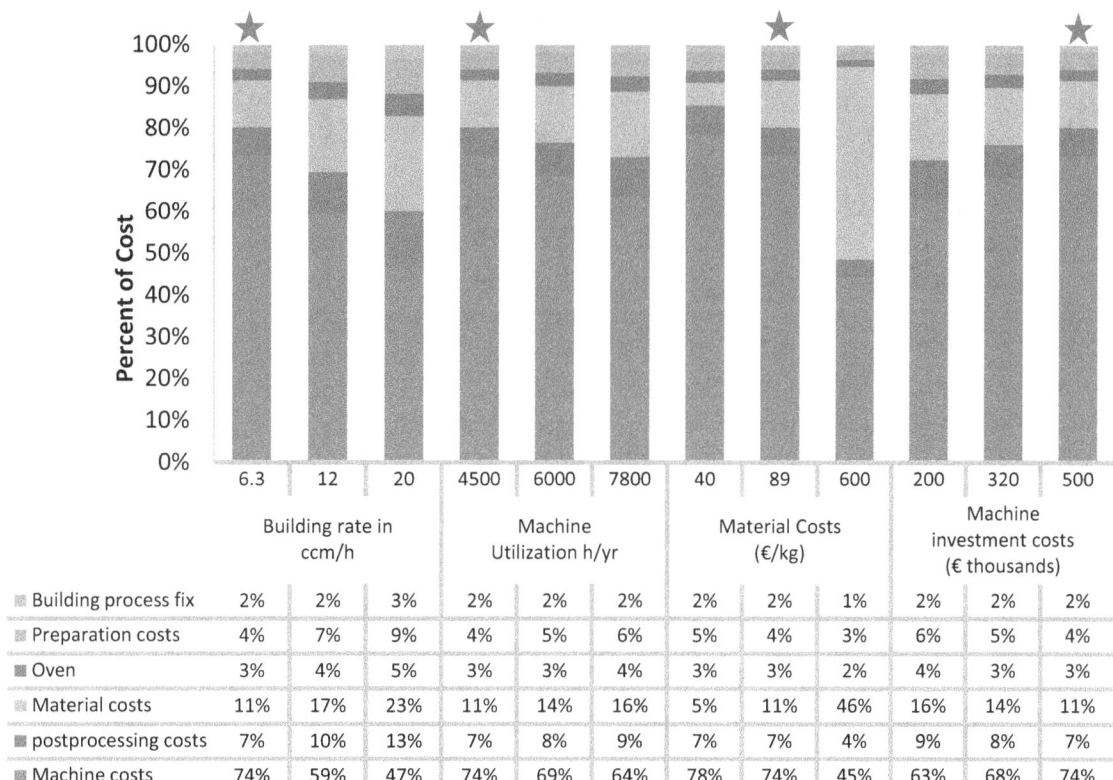

	6.3	12	20	4500	6000	7800	40	89	600	200	320	500
	\multicolumn{3}{c}{Building rate in ccm/h}										
Building process fix	2%	2%	3%	2%	2%	2%	2%	2%	1%	2%	2%	2%
Preparation costs	4%	7%	9%	4%	5%	6%	5%	4%	3%	6%	5%	4%
Oven	3%	4%	5%	3%	3%	4%	3%	3%	2%	4%	3%	3%
Material costs	11%	17%	23%	11%	14%	16%	5%	11%	46%	16%	14%	11%
postprocessing costs	7%	10%	13%	7%	8%	9%	7%	7%	4%	9%	8%	7%
Machine costs	74%	59%	47%	74%	69%	64%	78%	74%	45%	63%	68%	74%

Column groups: Building rate in ccm/h (6.3, 12, 20); Machine Utilization h/yr (4500, 6000, 7800); Material Costs (€/kg) (40, 89, 600); Machine investment costs (€ thousands) (200, 320, 500).

Source: Lindemann C., U. Jahnke, M. Moi, and R. Koch. "Analyzing Product Lifecycle Costs for a Better Understanding of Cost Drivers in Additive Manufacturing." Proceedings of the 2012 Solid Freeform Fabrication Symposium.
<http://utwired.engr.utexas.edu/lff/symposium/proceedingsArchive/pubs/Manuscripts/2012/2012-12-Lindemann.pdf>
Note: The orange star indicates the base model.

made of stainless steel. For this example, four cost factors are varied and the production quantity is a little less than 200 for the base case. This analysis provides insight into identifying the largest costs of additive manufacturing. The first cost factor that is varied is the building rate, which is the speed at which the additive manufacturing system operates. In this example, it is measured in cubic centimeters per hour. The second factor that is varied is the machine utilization measured as the number of hours per year that the machine is operated. The third factor is the material cost and the last factor is the machine investment costs, which include items related to housing, using, and maintaining the additive manufacturing system. Among other things, this includes energy costs, machine purchase, and associated labor costs to operate the system. The base model has a build rate of 6.3 ccm, a utilization of 4500 h/yr, a material cost of 89 €, and a machine investment cost of 500 000 €. For comparison, the base case is shown four times in the figure, with each one shown with a star. On average, the machine costs accounted for 62.9 % of the cost estimates in Figure 4.3 (note that the base case is only counted once in the average). This cost was the largest even when building rate was more than tripled and other factors were held constant. This cost was largest in all but one case, where material costs were increased to 600 €/kg. The second largest cost is the materials, which, on average, accounted for 18.0 % of the costs; however, it is important to note that this cost is likely to decrease as more suppliers enter the field.[48] Post processing, preparation, oven heating, and building process fix were approximately 8.4 %, 5.4 %, 3.3 %, and 1.9 %, respectively.

Plastic parts likely have a slightly different cost structure. A case study of a fluorescent lamp holder provides some insight. This case study examined two Electro Optical Systems that use selective laser sintering: P390 and P730. The P390 was more cost effective for this particular case study. The cost per part for this item was examined and revealed that for the P390, 58.7 % of the cost was machine cost, 9.9 % was machine operator cost, 30.4 % was material cost, and 1.0 % was assembly.[49]

For manufacturers, the cost advantage of additive manufacturing may vary. Typically, it is believed that this technology is competitive for low volume production. This can be illustrated in another case study of a landing gear assembly for a 1:5 scale model of the P180 Avant II by Piaggio Aero Industries S.p.A. As seen in Table 4.3, the per assembly cost of producing the landing gear using traditional manufacturing methods, in this case high pressure die casting, was 21.29 € plus 21 000 € divided by the lot size. The cost of additive manufacturing was 526.31 € per assembly; thus, below a lot size of 41 additive manufacturing is more cost effective. Above a lot size of 41 it was not cost effective. These cost estimates also illustrate how additive manufacturing does not follow traditional economies of scale, where large production runs reduce the per item cost; thus,

Table 4.3: High Pressure Die Cast Manufacturing Costs vs. Additive Manufacturing Costs (Selective Laser Sintering)

	Traditional Manufacturing (High Pressure Die Cast)	Additive Manufacturing (Selective Laser Sintering)
Material cost per part	2.59 €	25.81 €
Mould cost per part	21 000 €/N	-
Pre-processing cost per part	-	8.00 €
Processing cost per part	0.26 €	472.50 €
Post-processing cost per part	17.90 €	20.00 €
Linkages and assembly	0.54 €	-
TOTAL COST PER ASSEMBLY	21.29 €+21 000 €/N	526.31 €

Note: N is the lot size or the number of consecutive assemblies produced
Source: Atzeni, Eleonora, Luca Iuliano, and Alessandro Salmi. (2011) "On the Competitiveness of Additive Manufacturing for the Production of Metal Parts." Proceedings of the 9th International Conference on Advanced Manufacturing Systems and Technology.

---

[48] Lindemann C., U. Jahnke, M. Moi, and R. Koch. "Analyzing Product Lifecycle Costs for a Better Understanding of Cost Drivers in Additive Manufacturing." Proceedings of the 2012 Solid Freeform Fabrication Symposium.
<http://utwired.engr.utexas.edu/lff/symposium/proceedingsArchive/pubs/Manuscripts/2012/2012-12-Lindemann.pdf>
[49] Atzeni, Eleonora, Luca Iuliano, Paolo Minetola, and Alessandro Salmi. (2010) "Redesign and Cost Estimation of Rapid Manufactured Plastic Parts." Rapid Prototyping Journal. 16(5): 308-317.

each assembly produced using additive manufacturing costs the same regardless of how many are produced. The cost effectiveness of using additive manufacturing relies on a number of factors, including the complexity of the part, amount of material, and the volume of production.

5 Adoption and Diffusion of Additive Manufacturing

5.1 The Diffusion Process

Disseminating a new idea or innovation so that it is widely adopted can be difficult, even if it has obvious advantages. A common challenge for many is how to speed up the rate of diffusion of an innovation. *Diffusion*, for the purpose of this report, is defined as, "the spread of an innovation throughout a social system," while *adoption* is defined as, "the acceptance and continued use of a product, service, or idea."[50] The diffusion of new technologies or innovations tends to follow certain trends and the process is studied in several disciplines: economics, communications, sociology, and marketing.

There is both a diffusion model and an adoption model. The diffusion model is illustrated by the logistic S-curve that evaluates the time it takes for an innovation to be diffused into an industry.[51] Diffusion increases at an increasing rate up to time T_1, and then at a decreasing rate thereafter (see Figure 5.1).

Figure 5.1: The Logistical S-Curve Model of Diffusion

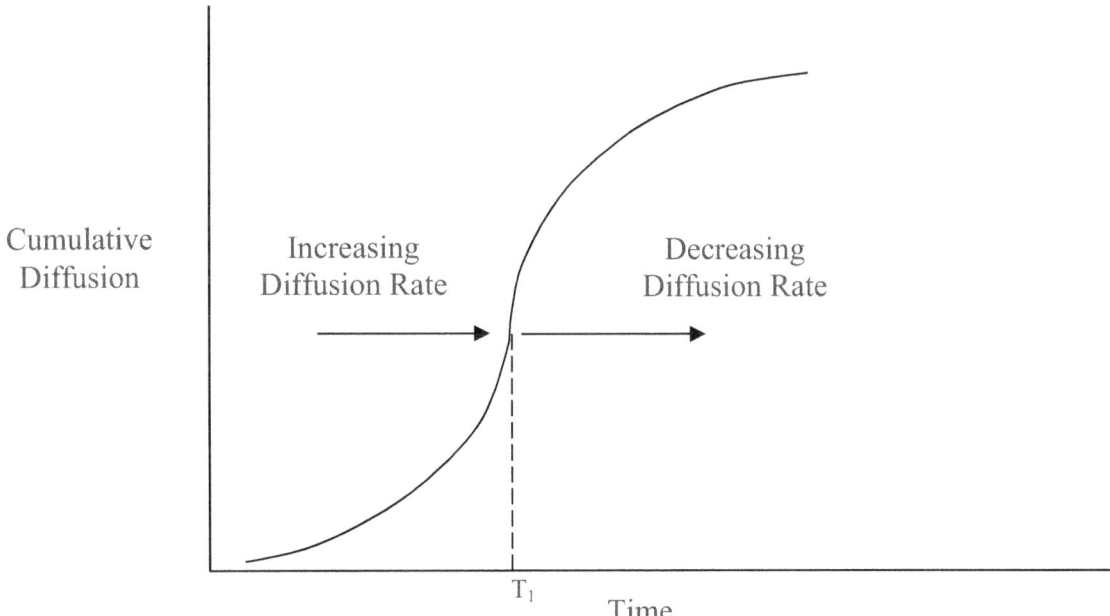

Modified from Rogers, E. M., (1995). *Diffusion of Innovations,* Fourth Edition, (New York: The Free Press, 1995), 258.

A simple logistic function may be defined by the following equation:

$$P(t) = \frac{1}{1 + e^{-t}}$$

---

[50] Koebel, C. Theodore, Maria Papadakis, Ed Hudson, Marilyn Cavell, *The Diffusion of Innovation in the Residential Building Industry*, PATH, p. 1.
[51] Ibid, p. 2.

Where *P* represents the population of adopters and *t* is time. The early growth is exponential and decays after 50 % of adopters are reached and *e* is Euler's number, the base of the natural system of logarithms.[52]

In connection with the diffusion model, the adoption model focuses on the decision process of the individual or firm. This model is connected with Everett Rogers' theory[53] that the S-curve is normally distributed (see Figure 5.2). Most adopters act in the midrange of the adoption period timeline because of information diffusion. This is where the adoption rate is the highest. At the "early adopters" stage in Figure 5.2, relatively little is known about the new technology and the number of adopters is low. At the stage of the "majority of adopters," a significant amount of information has been diffused. By the "late adopters" stage, there is little information remaining to be diffused. Each individual's adoption of the technology is equivalent to a "learning trial" in the system. Over time, adopter distributions follow a bell shaped curve.

Figure 5.2: Rogers' Model of Adoption (based on probability distribution)

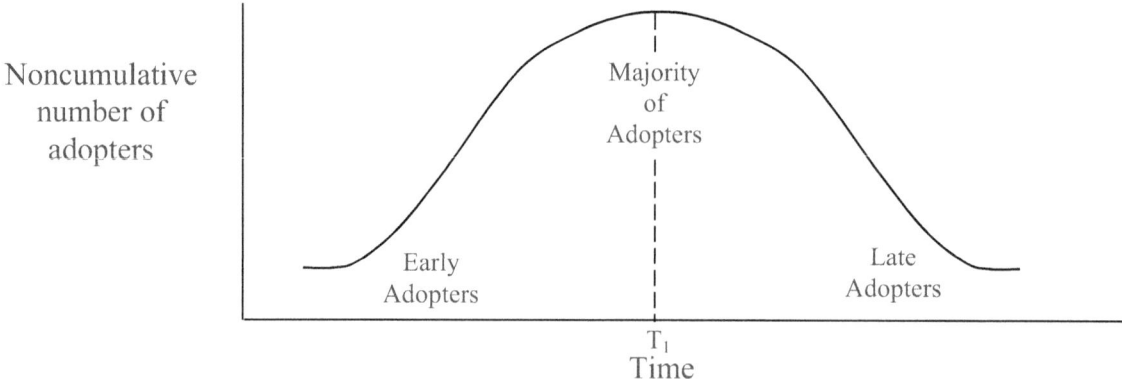

Modified from Rogers, E. M. (2003). *Diffusion of Innovations,* 5th Edition (New York: The Free Press, 2003), 111-114.

Larsen stresses three explanatory innovation diffusion concepts: (1) cohesion, (2) structural equivalence, and (3) thresholds.[54] *Cohesion* asserts that diffusion takes place by face-to-face contact between stakeholders, who are described as sharing a high degree of *homophily*; that is to say, they have a tendency to listen to people similar to themselves, whom they trust as friends. The stakeholder's logic behind listening to trusted friends relates to the risk and uncertainty of adopting new technology. *Structural equivalence* explains diffusion as a copycat approach. The decision to adopt is not based on sound judgment, but through fear and risk adversity. The last concept, *thresholds*, states that diffusion is a complex process that can be influenced by education, wealth, communication networks, and background. An innovation is not diffused over

---

[52] Vishwanath, Arun and George Barnett. *The Diffusion of Innovations.* (New York: Peter Lang, 2011).
[53] Rogers, E. M. (2003). *Diffusion of Innovations,* Fourth Edition (New York: The Free Press, 2003), p. 111-114.
[54] Larsen, Graeme D., "Horses for Courses: Relating Innovation Diffusion Concepts to the Stages of the Diffusion Process," *Construction Management and Economics*, Vol 23, October 2005, p. 787-792.

homogenous people, but between diverse individuals with different backgrounds. According to the concept of thresholds, a stakeholder's decision to adopt a new technology is interconnected with other stakeholders.[55]

5.2 Factors of Diffusion

Some innovations, such as cellular phones, only take a few years to reach widespread adoption, while others can take decades. Characteristics of innovations can provide some explanation for this difference. Rogers identifies five primary characteristics as seen in Figure 5.3: relative advantage, compatibility, complexity, trialability, and observability. The relative advantage is the extent that an innovation is perceived to be better than the current or previous idea. Compatibility is the extent that a new innovation is consistent with current values and needs. Innovations that are compatible with current norms and needs are likely to be adopted more rapidly than one that is not compatible. Complexity refers to the perception of how complicated a new innovation is to understand and use. Increased complexity slows the adoption of a new innovation. Trialability is the extent that a new innovation may be tested before fully adopting it. Observability is the extent that the use and results of a new innovation can be seen by would-be adopters.

The type of innovation decision is also a factor in the rate of adoption. Optional innovation decisions are decisions made by individuals independent of other members of a system; thus, the individual is the main unit of decision making. Collective innovation decisions are those decisions that are made by consensus among members of a system. Authority innovation decisions are those decisions to adopt or reject an innovation by a

Figure 5.3: Variables Determining the Rate of Adoption of Innovations

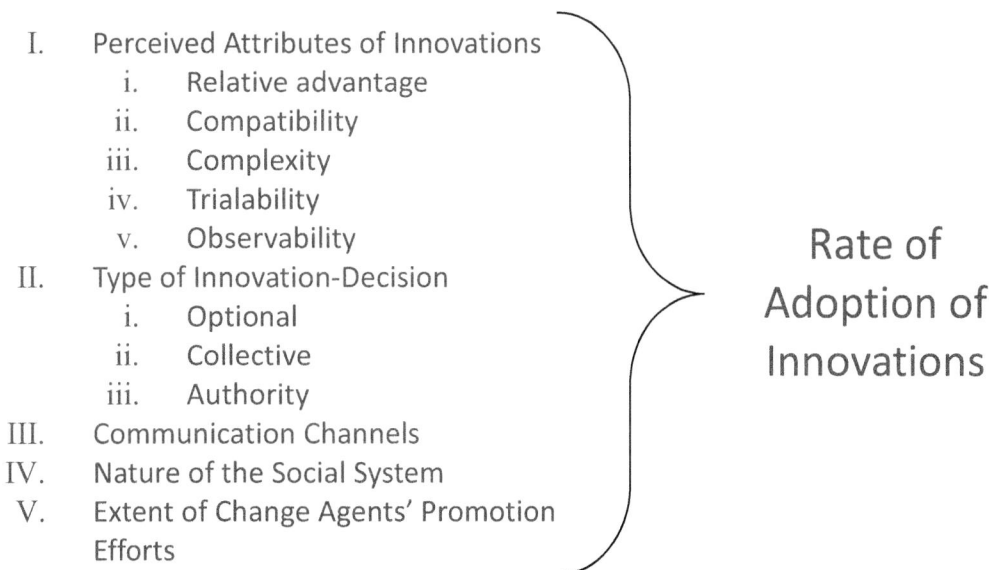

Rogers, E. M. (1995). *Diffusion of Innovations,* 5th Edition (New York: The Free Press, 2003), 222.

---

[55] Larsen, Graeme D., "Horses for Courses: Relating Innovation Diffusion Concepts to the Stages of the Diffusion Process," *Construction Management and Economics*, Vol 23, October 2005, p. 787-792.

select few individuals who maintain power, status, or technical expertise. For example, a chief executive officer (CEO) who decides that all employees will wear a suit would be an authority decision.

A communication channel, as referred to in Figure 5.3, is the means by which individuals communicate concerning an innovation. These might include the evaluation of an innovation by a peer or a review by an expert. One-on-one and other communications often take place within a social system. Communication may also occur through mass media. Communication channels are important in determining the diffusion of an innovation; however, it often requires in-depth investigation to understand these channels.

The nature of the social system, such as its norms and interconnectedness, is also an important factor in the diffusion of an innovation. This includes the system's culture, but also includes the network of connections between potential adopters. This can be a significant factor in the diffusion of a technology, especially in the case where the preferred communication channel is one-on-one interaction. Similar to the communication channels, the nature of the social system is an important factor, but this type of information is not well documented. Additional research may be needed to develop a full understanding of both the social system and relevant communication channels.

The last variable is the change agent. Both public and private organizations strive to change the marketplace. Many entities provide incentives or subsidies in order to speed up the rate of adoption of innovations. For example, the federal government often creates incentives for individuals or businesses to adopt more environmentally friendly products such as energy efficient lighting. Other events, organizations, people, or items also act as a catalyst for change in an industry.

5.3 Diffusion of Additive Manufacturing

Globally, 6494 industrial additive manufacturing systems were deployed in 2011 with a cumulative total of 49 035 systems being deployed between 1988 and 2011. Of these, 18 780 were deployed in the U.S. The growth in the cumulative number of additive manufacturing systems in the U.S. between 2010 and 2011 was 15.3 %.[56]

The status of some of the variables that affect the adoption of additive manufacturing technologies can be observed through existing articles and texts; however, many issues cannot be substantiated without gathering additional data. Surveys can often be used to assess a producer or user's opinion of a new technology, but this is often a resource intensive process. Using the number of domestic unit sales[57], the growth in sales can be fitted using least squares criterion to an exponential curve that represents the traditional

---

[56] Wohlers, Terry. "Wohlers Report 2012: Additive Manufacturing and 3D Printing State of the Industry." Wohlers Associates, Inc. 2012.
[57] Wohlers, Terry. "Wohlers Report 2012: Additive Manufacturing and 3D Printing State of the Industry." Wohlers Associates, Inc. 2012.

logistic S-curve of technology diffusion. The most widely accepted model of technology diffusion was presented by Mansfield[58]:

$$p(t) = \frac{1}{1 + e^{\alpha - \beta t}}$$

Where

$p(t)$ = the proportion of potential users who have adopted the new technology by time t
α = Location parameter
β = Shape parameter ($\beta > 0$)

In order to examine additive manufacturing, it is assumed that the proportion of potential units sold by time t follows a similar path as the proportion of potential users who have adopted the new technology by time t. In order to examine shipments in the industry, it is assumed that an additive manufacturing unit represents a fixed proportion of the total revenue; thus, revenue will grow similarly to unit sales. The proportion used was calculated from 2011 data. The variables α and β are estimated using regression on the cumulative annual sales of additive manufacturing systems in the U.S. between 1988 and 2011. U.S. system sales are estimated as a proportion of global sales. This method provides some insight into the current trend in the adoption of additive manufacturing technology. Unfortunately, there is little insight into the total market saturation level for additive manufacturing; that is, there is not a good sense of what percent of the relevant manufacturing industries (shown in Table 4.1) will produce parts using additive manufacturing technologies versus conventional technologies. In order to address this issue, a modified version of Mansfield's model is adopted from Chapman[59]:

$$p(t) = \frac{\eta}{1 + e^{\alpha - \beta t}}$$

Where
η = market saturation level

Because η is unknown, it is varied between 0.15 % and 100 % of the relevant manufacturing shipments, as seen in Table 5.1. The 0.15 % is derived from Wohlers estimate that the 2011 sales revenue represents 8 % market penetration, which equates to $3.1 billion in market opportunity and 0.15 % market saturation. At this level, additive manufacturing is forecasted to reach 50 % market potential in 2018 and 100 % in 2045, as seen in the table. A more likely scenario seems to be that additive manufacturing would have between 5 % and 35 % market saturation. At these levels, additive manufacturing would reach 50 % of market potential between 2031 and 2038 while

---

[58] Mansfield, Edwin. *Innovation, Technology and the Economy: Selected Essays of Edwin Mansfield.* Economists of the Twentieth Century Series (Brookfield, VT: 1995, E. Elgar).
[59] Chapman, Robert. "Benefits and Costs of Research: A Case Study of Construction Systems Integration and Automation Technologies in Commercial Buildings." NISTIR 6763. December 2001. National Institute of Standards and Technology.

reaching 100 % between 2058 and 2065, as seen in Table 5.1. The industry would reach $50 billion between 2029 and 2031 while reaching $100 billion between 2031 and 2044. As illustrated in Figure 5.4, it is likely that additive manufacturing is at the far left tail of the diffusion curve, making it difficult to forecast the future trends; thus, some caution

Table 5.1: Forecasts of U.S. Additive Manufacturing Shipments by Varying Market Potential

Market Potential of Relevant Manufacturing (percent of shipments)	Market Potential, Shipments ($billions 2011)	Approximate Year 100% of Market Potential Reached	Approximate Year 50% of Market Potential Reached	Approximate Year $100 Billion in Shipments is Reached	Approximate Year $50 Billion in Shipments is Reached	R^2
100.00	$2 058.9	2069	2042	2031	2028	0.948
75.00	$1 544.2	2068	2041	2031	2028	0.948
50.00	$1 029.5	2067	2039	2031	2029	0.948
45.00	$926.5	2066	2039	2031	2029	0.948
40.00	$823.6	2066	2038	2031	2029	0.948
35.00	$720.6	2065	2038	2031	2029	0.948
30.00	$617.7	2065	2037	2031	2029	0.948
25.00	$514.7	2064	2037	2032	2029	0.948
20.00	$411.8	2063	2036	2032	2029	0.948
15.00	$308.8	2062	2035	2032	2029	0.948
10.00	$205.9	2061	2033	2033	2029	0.948
5.00	$102.9	2058	2031	2044	2031	0.948
1.00	$20.6	2052	2025	-	-	0.949
0.50	$10.3	2050	2023	-	-	0.949
0.15	$3.1	2045	2018	-	-	0.950

should be used when interpreting this forecast. The figure illustrates the diffusion at each market saturation level presented in Table 5.1 with the exception of the 0.50 % and 0.15 % levels, as they are too small to be included in this graph.

5.3.1 Perceived Attributes of Innovation

Relative Advantage: The relative advantage of adopting additive manufacturing varies from industry to industry and is likely to increase over time as the technology advances. The per-unit cost of additive manufacturing appears to be a significant barrier for many would-be adopters. For some, the benefits outweigh the costs. For instance, lighter transportation equipment can significantly reduce costs for end users; thus, they might be willing to pay higher upfront costs to purchase lighter equipment made using additive manufacturing technologies. For others, however, the benefits of products made using this technology may not justify the higher costs for producers or end users. One possible challenge that could develop is communicating and convincing the end user of the benefits of a product made using additive manufacturing. For instance, this technology may allow for the design of a longer lasting product; however, the end user is only willing to pay for the additional costs of production if they are aware of and convinced of the benefits.

One of the primary beneficiaries of additive manufacturing is the end user; thus, their role in persuading manufacturers to adopt additive manufacturing technology is a significant one. On the other hand, manufacturers may need to differentiate products made using additive manufacturing technology by indicating the benefits to the end user; otherwise, costumers may not be willing to pay the costs for these products.

Compatibility: The limited size of the products that can be produced affects the compatibility of additive manufacturing for some manufactured products. Transportation

Figure 5.4: Forecasts of U.S. Additive Manufacturing Shipments, by Varying Market Saturation Levels

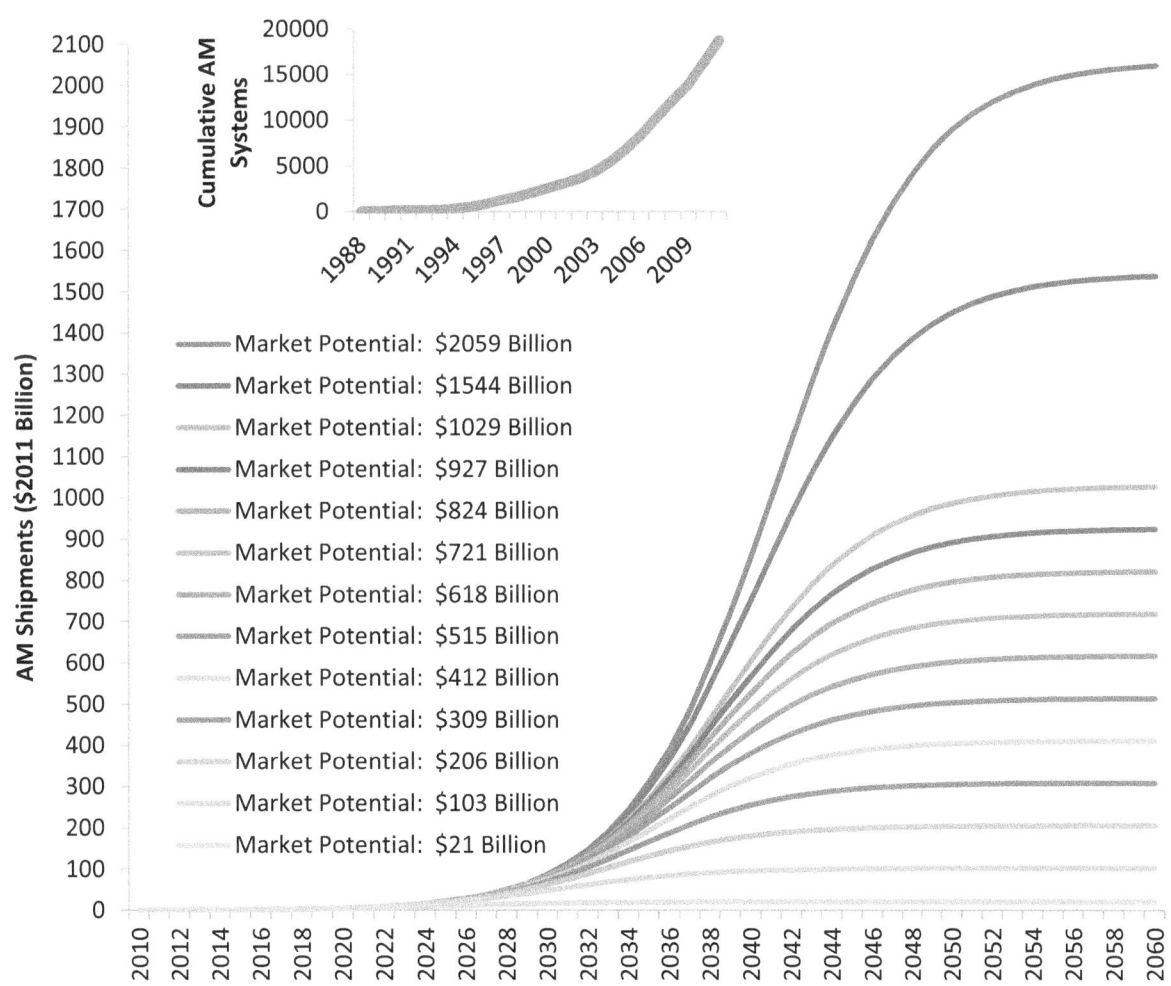

equipment, for instance, involves large components that may be difficult to produce using additive manufacturing technology.

Complexity, Trialability and Observability: Additive manufacturing systems can be costly; however, these systems are seemingly easy to illustrate and a significant amount of literature is available on them. Currently, there are journals and conferences that discuss this technology extensively. One challenge that seems to persist is cost categorization and analysis. This prevents a prospective manufacturer from observing the costs and benefits from adopting this technology. A number of developments have been made on this front; however, no model meets all criteria adequately. There is a need to bring together the strengths of existing cost models into one standardized model.[60] This would allow would-be adopters to understand the benefits and costs more adequately.

5.3.2 Change Agents

The last factor involves the efforts of change agents. These entities can be individuals, events, organizations, or some other entity that acts as a catalyst for change. They often accelerate the realization of benefits, reduce costs, and/or increase benefits of some trend in society or the economy. This change can often occur through research and collaboration efforts. For additive manufacturing, there are a number of organizations that strive to advance the current status. One newly created organization is the National Additive Manufacturing Innovation Institute (NAMII), which was formally established in 2012 with an initial $30 million in federal funding matched by $40 million from a consortium of companies, universities, colleges, and non-profit organizations. The single focus of NAMII is to "accelerate additive manufacturing technologies to the U.S. manufacturing sector and increase domestic manufacturing competitiveness."[61] Likewise, the Additive Manufacturing Consortium (AMC) was launched by EWI. The mission of the AMC is to "bring together a diverse group of practitioners and stakeholders that together accelerate the innovation in AM technologies to move them into the mainstream of manufacturing technology from their present emerging position."[62]

---

[60] Lindemann C., U. Jahnke, M. Moi, and R. Koch. "Analyzing Product Lifecycle Costs for a Better Understanding of Cost Drivers in Additive Manufacturing." Proceedings of the 2012 Solid Freeform Fabrication Symposium.
<http://utwired.engr.utexas.edu/lff/symposium/proceedingsArchive/pubs/Manuscripts/2012/2012-12-Lindemann.pdf>
[61] National Additive Manufacturing Innovation Institute. <http://namii.org/>
[62] EWI. Additive Manufacturing Consortium. < http://ewi.org/additive-manufacturing-consortium/>

6 Opportunities for Change Agents

Metrics used to discuss national industries often involve examining the amount of research being conducted, factors that impact the industry, or the size of the industry. The primary purpose of investing resources into manufacturing activities, however, is to generate a benefit or return on investment. Arguably, those countries that exceed the U.S. in per capita size and benefits per unit of input, such as compensation per hour, have an industry that is more successful at the main objective of investing resources in manufacturing. The general purpose of an industry change agent is to create a net increase in the return on investment for stakeholders. For additive manufacturing, this might be accomplished by reducing costs, accelerating the realization of benefits, or increasing the net benefits as illustrated in the larger graph illustrated in Figure 6.1. These changes result in an increase in the marginal return on investment as illustrated in the

Figure 6.1: Impact of Change Agents on the Net Benefits and Return on Investment for Additive Manufactured Products

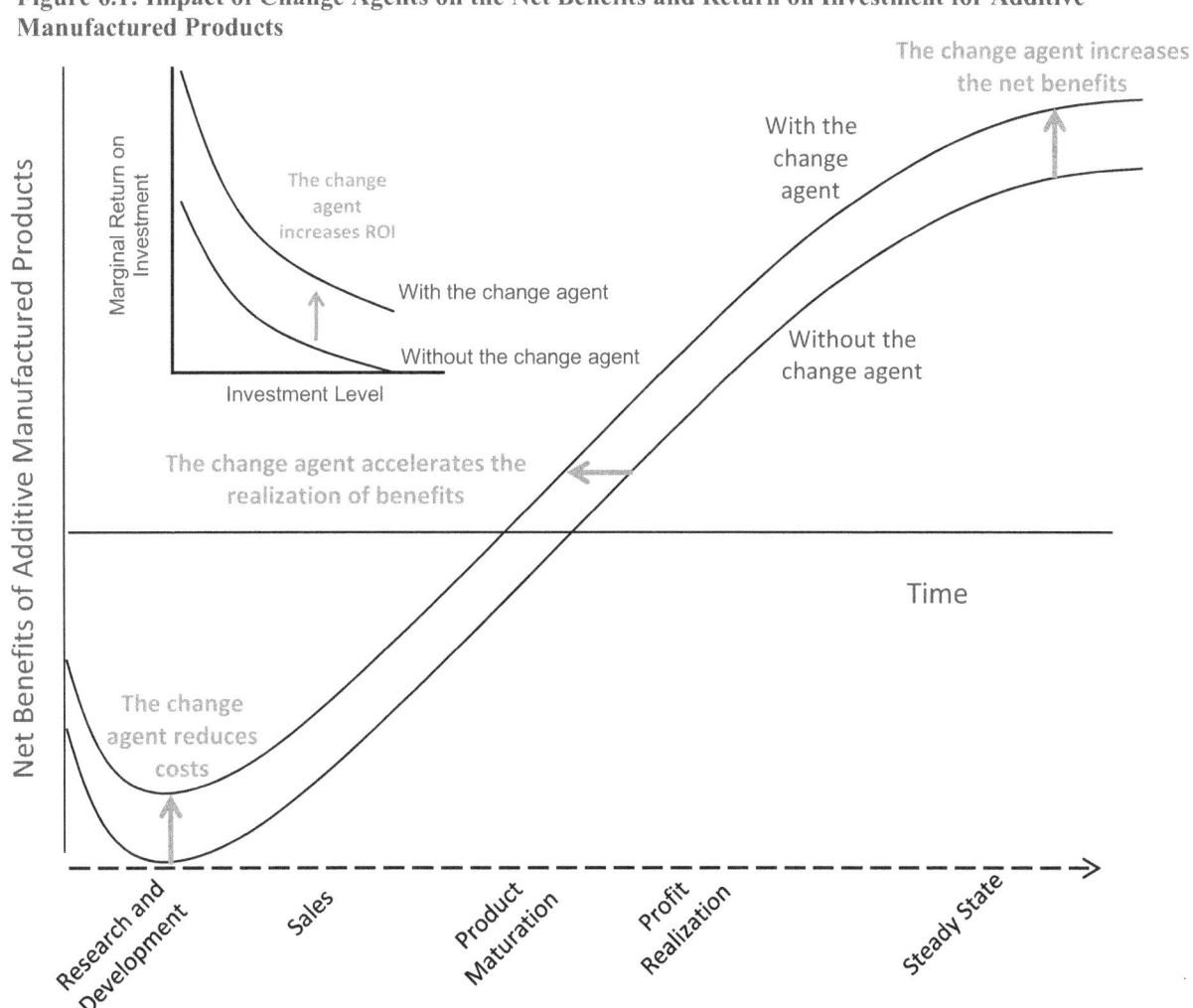

Modified from Gallaher, Michael P., Thomas Phelps and Alan C. O'Connor. Planning Report 02-5: Economic Impact Assessment of the International Standard for the Exchange of Product Model Data (STEP) in Transportation Equipment Industries. RTI International and the National Institute of Standards and Technology. December 2002: 5-4.

smaller graph in the figure.

Generally, change agents want to maximize their impact for the amount of resources allotted to them; that is, they want the "biggest bang for the buck." Investment in any particular change agent effort, traditionally, has decreasing returns to scale; that is, every additional dollar of investment has a little less impact than the previous dollar. Since a change agent wants to maximize their impact, it would want to allocate its funding in projects such that each dollar of investment has the maximum return possible. For instance, Figure 6.2 provides an illustration of five possible investments for a change agent with a budget constraint. The investments are referred to as efforts A through E. To maximize its impact, a change agent would first invest in Effort A. As it invests more and more in Effort A, it moves from the left to the right along the marginal return on

Figure 6.2: Illustration of the Optimal use of Change Agent Funding for Six Alternative Investments

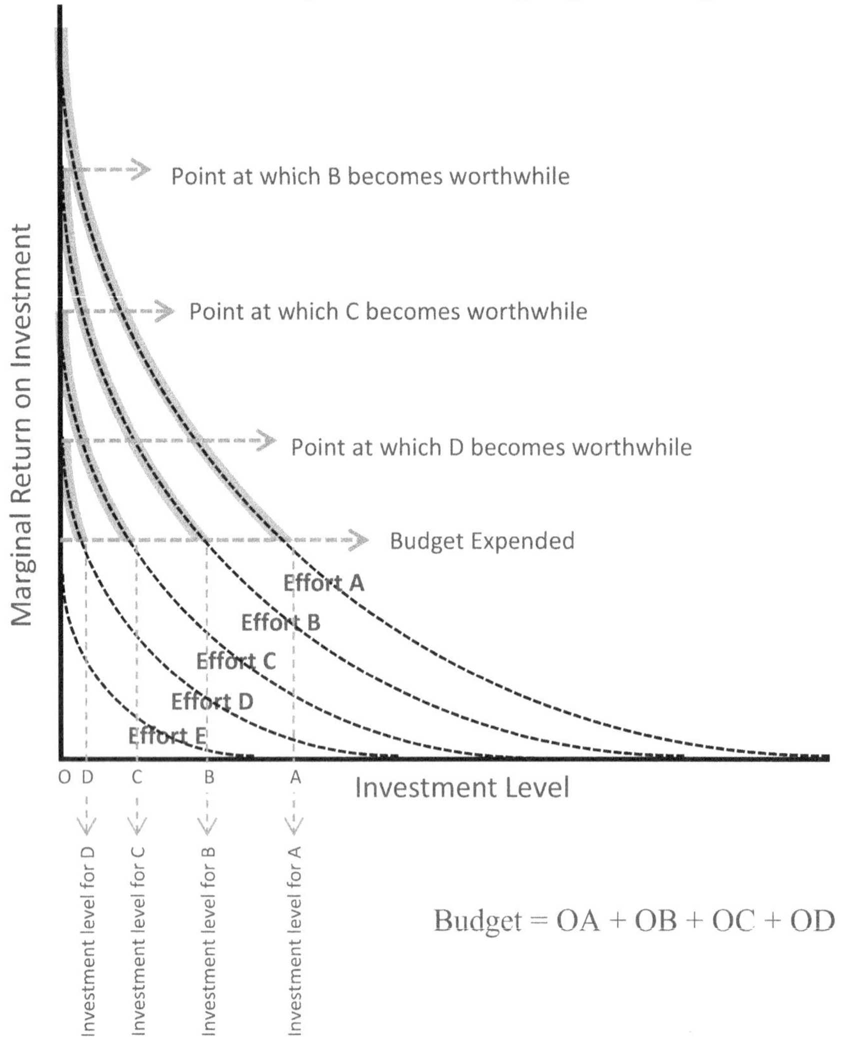

$$\text{Budget} = OA + OB + OC + OD$$

Note: The green lines represent the investment for each effort.

investment line for the change agent. The agent would invest to the point where the marginal return on investment for its next dollar invested equals that of Effort B, which is referred to in the figure as the "Point at which B becomes worthwhile." At this point, there is some indifference to investing in A or B because they have the same marginal return on investment; however, as one invests in either A or B the return on investment in that effort decreases making the alternative more appealing. Therefore, the change agent would invest in both A and B or alternate between the two until the point at which the next effort becomes worthwhile. It would continue to do this until its entire budget is expended. In this example, effort E goes unfunded while efforts A through D are funded to where the bottom of each corresponding green line stops; thus, the total investment is the sum of the investment level for Effort A, B, C, and D. It is important to note, however, that not all of the costs and benefits of the manufacturing industry are able to be measured nor are the impact of the efforts of change agents; therefore, identifying the optimal use of funding can be rather problematic.

Change agents for the additive manufacturing industry can focus their efforts on three primary areas to advance this technology: cost reduction, accelerating the realization of benefits, and increasing the benefits of additive manufacturing. The costs include any of the investments made by the stakeholders listed in Table 3.2. These include the owners, employees, suppliers, and end users among others. The producer costs of additive manufacturing tend to be broken into preparation, materials, machine utilization, and post processing. As seen in some case studies in Section 4.3, the largest cost tends to be the machine operation cost followed by the material cost. The time it takes to produce a product may be a significant factor in the machine utilization cost. Since these two costs are the largest, there is a potentially high marginal return on investment for change agents that focus on reducing these costs; that is, focusing on these items may result in a higher return on investment for some change agents. However, examining these issues in detail provides some challenge as there is not a standard cost categorization. This prevents change agents from precisely identifying the major costs of this technology. A number of developments have been made on this front; however, no model meets all criteria adequately. There is a need to bring together the strengths of existing cost models into one standardized model.[63] This might be another area that has a high return on investment for change agents.

As previously discussed, a major benefit of this technology is in the area of product design and how it allows the production of nearly any complexity of geometry without the need for tooling. Additionally, the complexity does not impact the cost in the same way that it does for conventional manufacturing. This technology eliminates many of the restrictions of 'Design for Manufacture and Assembly' opening a new realm of possibilities for new customized products at an affordable price point. To some degree, the success of this technology will rely on taking advantage of this benefit. In order to

---

[63] Lindemann C., U. Jahnke, M. Moi, and R. Koch. "Analyzing Product Lifecycle Costs for a Better Understanding of Cost Drivers in Additive Manufacturing." Proceedings of the 2012 Solid Freeform Fabrication Symposium.
<http://utwired.engr.utexas.edu/lff/symposium/proceedingsArchive/pubs/Manuscripts/2012/2012-12-Lindemann.pdf>

achieve this, the products must meet quality and reliability standards and there must be testing standards in place to verify their performance. For instance, the U.S. Federal Aviation Regulations have strict regulations for material performance related to fatigue, creep, flammability, and toxicity. Manufacturers rely on standards in materials and processes to ensure the performance of their products.[64] The dissimilarities between conventional manufacturing processes and those of additive manufacturing are likely to require modifications to current performance validation processes.[65] Standards and codes organizations will likely play a significant role in facilitating the adoption of additive manufacturing technology.

Although this technology can produce nearly any complexity of geometry, it is limited in the size of the components that can be constructed. Expanding the size while maintaining a reasonable price point is likely to increase the rate at which this technology is adopted and expand the market opportunity. Additionally, the quality of the product is a limiting factor. Materials or surface finish, for instance, can often be inadequate for parts and components.

---

[64] National Academy of Engineering. "Frontiers of Engineering 2011: Reports on Leading-Edge Engineering from the 2011 Symposium." In National Academy of Engineering's 2011 U.S. Frontiers of Engineering Symposium. Mountain View, CA. 2012

[65] Scott, Justin, Nayanee Gupta, Christopher Weber, Sherrica Newsome, Terry Wohlers, and Tim Caffrey. "Additive Manufacturing: Status and Opportunities", March 2012. <https://www.ida.org/stpi/occasionalpapers/papers/AM3D_33012_Final.pdf>

7 Conclusion

There is a general concern that the U.S. manufacturing industry has lost competitiveness with other nations; however, industry data suggests that the U.S. still maintains a prominent position. Additive manufacturing may provide an important opportunity for advancing U.S. manufacturing while maintaining and advancing U.S. innovation. The U.S. is currently a major user of additive manufacturing technology and the primary producer of additive manufacturing systems. Globally, an estimated $642.6 million in revenue was collected for additive manufactured goods, with the U.S. accounting for an estimated $246.1 million or 38.3 % of global production in 2011. Approximately 62.8% of all commercial/industrial units sold in 2011 were made by the top three producers of additive manufacturing systems: Stratasys, Z Corporation, and 3D Systems based out of the U.S. Approximately 64.4% of all systems were made by companies based in the U.S. If additive manufacturing has a saturation level between 5 % and 35 % of the relevant sectors, it is forecasted that it might reach 50 % of market potential between 2031 and 2038, while reaching 100 % between 2058 and 2065, as seen in Table 5.1. The industry would reach $50 billion between 2029 and 2031, while reaching $100 billion between 2031 and 2044. Since it is likely that additive manufacturing is at the far left tail of the diffusion curve, making it difficult to forecast the future trends, some caution should be used when interpreting these estimates.

Change agents for the additive manufacturing industry can focus their efforts on three primary areas: costs, rate at which benefits are realized, or the benefits of additive manufacturing. Costs have been identified as being a significant factor in whether producers adopt additive manufacturing technologies. Hopkinson[66] estimates that machine costs range between 50 % and 75 % of total cost, materials range between 20 % and 40 %, and labor ranges between 5 % and 30 %. Reducing these costs may have a significant impact on the adoption of additive manufacturing technologies. Additionally, quality, performance validation, and expanding size capabilities are likely to also have significant impacts.

---

[66] Hopkinson, Neil, "Production Economics of Rapid Manufacture." In Hopkinson, Neil, Richard Hague, and Philip Dickens. *Rapid Manufacturing*. (Hoboken, NJ: John Wiley & Sons, 2006).

Literature Cited

Atzeni, Eleonora, Luca Iuliano, Paolo Minetola, and Alessandro Salmi. (2010) "Redesign and Cost Estimation of Rapid Manufactured Plastic Parts." Rapid Prototyping Journal. 16(5): 308-317.

Boothroyd, Geoffrey, Peter Dewhurst, and Winston Knight. Product Design for Manufacture and Assembly. (New York: Marcel Dekker, Inc, 2009).

Bourell, David L., Ming C. Leu, and David W. Rosen. "Roadmap for Additive Manufacturing: Identifying the Future of Freeform Processing." University of Texas. <http://wohlersassociates.com/roadmap2009.html>

Chapman, Robert. "Benefits and Costs of Research: A Case Study of Construction Systems Integration and Automation Technologies in Commercial Buildings." NISTIR 6763. December 2001. National Institute of Standards and Technology.

Davidson, Adam. "Making It in America." The Atlantic. January/February (2012). <http://www.theatlantic.com/magazine/archive/2012/01/making-it-in-america/8844/?single_page=true>

Davidson, Adam. "The Transformation of American Factory Jobs, In One Company." NPR. January 13, 2012. <http://www.npr.org/blogs/money/2012/01/13/145039131/the-transformation-of-american-factory-jobs-in-one-company?ft=1&f=100>

Economist. "Printing Body Parts: Making a Bit of Me." <http://www.economist.com/node/15543683>

Gausemeier, Jurgen, Niklas Echterhoff, Martin Kokoschika, and Marina Wall. "Thinking Ahead the Future of Additive Manufacturing – Future Applications." University of Paderborn, Direct Manufacturing Research Center.

Gibson, Ian, David Rosen, and Brent Stucker. Additive Manufacturing Technologies. Springer: New York, 2010. 47-50

GizMag. "3D Bio-printer to Create Arteries and Organs." <http://www.gizmag.com/3d-bio-printer/13609/>

Greenwald, Bruce C.N. and Judd Kahn. Globalization: The Irrational Fear that Someone in China will Take Your Job. (Hoboken, NJ: John Wiley & Sons 2009).

Hopkinson, Neil, "Production Economics of Rapid Manufacture." In Hopkinson, Neil, Richard Hague, and Philip Dickens. Rapid Manufacturing. (Hoboken, NJ: John Wiley & Sons, 2006). 147-157.

Horowitz, Karen J. and Mark A. Planting. Concepts and Methods of the U.S. Input-Output Accounts. Bureau of Economic Analysis. 2006.

Koebel, C. Theodore, Maria Papadakis, Ed Hudson, Marilyn Cavell, The Diffusion of Innovation in the Residential Building Industry, PATH, p. 1.

Krugman, Paul R. "Competitiveness, A Dangerous Obsession." Foreign Affairs. Vol 73. Num 2. March/April (1994): 28-44.

Krugman, Paul R. "Making Sense of the Competitiveness Debate." Oxford Review of Economic Policy. Vol 12, no. 3 (1996): 17-25. Paul Krugman won the 2008 Nobel Memorial Prize in Economic Sciences for his work on international trade and economic geography.

Larsen, Graeme D., "Horses for Courses: Relating Innovation Diffusion Concepts to the Stages of the Diffusion Process," Construction Management and Economics, Vol 23, October 2005, p. 787-792.

Lindemann C., U. Jahnke, M. Moi, and R. Koch. "Analyzing Product Lifecycle Costs for a Better Understanding of Cost Drivers in Additive Manufacturing." Proceedings of the 2012 Solid Freeform Fabrication Symposium.
<http://utwired.engr.utexas.edu/lff/symposium/proceedingsArchive/pubs/Manuscripts/2012/2012-12-Lindemann.pdf>

Mansfield, Edwin. Innovation, Technology and the Economy: Selected Essays of Edwin Mansfield. Economists of the Twentieth Century Series (Brookfield, VT: 1995, E. Elgar).

Mansour, S., Richard Hague. (2003) "Impact of Rapid Manufacturing on Design for Manufacture for Injection Molding." Proceedings of the Institution of Mechanical Engineers, Part B: Journal of Engineering Manufacture.

McKinsey&Company. "Manufacturing the Future: The Next Era of Global Growth and Innovation." November 2012.
<http://www.mckinsey.com/insights/mgi/research/productivity_competitiveness_and_growth/the_future_of_manufacturing>

National Academy of Engineering. "Frontiers of Engineering 2011: Reports on Leading-Edge Engineering from the 2011 Symposium." In National Academy of Engineering's 2011 U.S. Frontiers of Engineering Symposium. Mountain View, CA. 2012

National Institute of Standards and Technology. "Roadmapping Workshop: Measurement Science for Metal-Based Additive Manufacturing." <http://events.energetics.com/nist-additivemfgworkshop/index.html>

National Science Foundation. "Asia's Rising Science and Technology Strength." May 2007. <http://www.nsf.gov/statistics/nsf07319/>

OECD (2012), OECD Science, Technology and Industry Outlook 2012, OECD Publishing. <http://dx.doi.org/10.1787/sti_outlook-2012-en>

Porter, Michael E. "Building the Microeconomic Foundations of Prosperity: Findings from the Business Competitiveness Index." In Porter, Michael E., Klaus Schwab, Xavier Sala-i-Martin, and Augusta Lopez-Claros. The Global Competitiveness Report 2003-2004. (New York: Oxford University Press, 2004).

Porter, Michael E. The Competitive Advantage of Nations. 1st ed. (New York: The Free Press, 1990).

Rogers, E. M. (2003). Diffusion of Innovations, Fourth Edition (New York: The Free Press, 2003), p. 111-114.

Scott, Justin, Nayanee Gupta, Christopher Weber, Sherrica Newsome, Terry Wohlers, and Tim Caffrey. "Additive Manufacturing: Status and Opportunities", March 2012.
<https://www.ida.org/stpi/occasionalpapers/papers/AM3D_33012_Final.pdf>

Sirkin, Harold L. "Made in the USA Still Means Something." Bloomberg Businessweek. April 10, 2009. <http://www.businessweek.com/managing/content/apr2009/ca20090410_054122.htm>

Slaughter, Matthew J. "How U.S. Multinational Companies Strengthen the U.S. Economy." United States Council for International Business. (March 2010).
<http://www.uscib.org/docs/foundation_multinationals.pdf>

Tassey Gregory. "Rationales and Mechanisms for Revitalizing U.S. Manufacturing R&D Strategies." Journal of Technology Transfer. 35 (2010): 283-333.

Thomas, Douglas S. "The Current State and Recent Trends of the U.S. Manufacturing Industry", NIST Special Publication 1142. December 2012. <http://www.nist.gov/manuscript-publication-search.cfm?pub_id=912933>

Thomas, Douglas. "National Industry Performance Metrics: A Case Study of U.S. Manufacturing." National Institute of Standards and Technology. White paper. Available upon request.

Thomson Reuters. "Top 100 Global Innovators, 2011." <http://www.top100innovators.com/overview>

Triadic patent families are defined at the OECD as a set of patents taken at the European Patent Office, Japanese Patent Office, and U.S. Patent and Trademark Office that share one or more priorities.

Vishwanath, Arun and George Barnett. The Diffusion of Innovations. (New York: Peter Lang, 2011).

West, Karl. "Melted Metal Cuts Plane's Fuel Bill." The Sunday Times. Sunday 13 February 2011. <http://www.thesundaytimes.co.uk/sto/business/energy_and_environment/article547163.ece>

Wohlers, Terry. "Wohlers Report 2012: Additive Manufacturing and 3D Printing State of the Industry." Wohlers Associates, Inc. 2012: 130.

World Economic Forum. The Global Competitiveness Report. 2010-2011. <http://www3.weforum.org/docs/WEF_GlobalCompetitivenessReport_2010-11.pdf>

Appendix A: Schematic Data Map

The Annual Survey of Manufactures (ASM) is conducted every year except for years ending in 2 or 7 when the Economic Census is conducted. The ASM provides statistics on employment, payroll, supplemental labor costs, cost of materials consumed, operating expenses, value of shipments, value added, fuels and energy used, and inventories. It uses a sample survey of approximately 50 000 establishments with new samples selected at 5-year intervals. An establishment is an economic unit—business or industrial—at a single physical location where business is conducted or where services or industrial operations are performed. The ASM data allows the examination of multiple factors (value added, payroll, energy use, and more) of manufacturing at a detailed subsector level. The Economic Census, used for years ending in 2 or 7, is a survey of all employer establishments in the U.S. that has been taken as an integrated program at 5-year intervals since 1967. Both the ASM and the Economic Census use NAICS classification; however, prior to NAICS the Standard Industrial Classification system was used. Table A.1 contains items from the Annual Survey of Manufactures. The color scheme matches that of the color scheme in the manufacturing supply chains presented previously in this report.

Each supply chain item is calculated for the NAICS codes listed in Table 4.1 and added together by the categories listed in the table using data from the Annual Survey of Manufactures seen in Table A.2. The values for additive manufacturing seen in Table A.3 are calculated by assuming that the ratio of each supply chain item to the total value of shipments from the data in Table A.2 is the same for additive manufacturing. The ratios are then applied to data in the 2012 Wohlers Report. These assumptions have significant implications for precision; however, they are the best estimates available.

Table A.1: Supply Chain Components

ASM Data Item	Schematic name
Number of employees	Payroll, Benefits, and employment
Annual payroll	Payroll, Benefits, and employment
Total fringe benefits	Payroll, Benefits, and employment
Employer's cost for health insurance	
Employer's cost for defined benefit pension plans	
Employer's cost for defined contribution plans	
Employer's cost for other fringe benefits	
Production workers avg per year	
Production workers hours (1,000)	
Production workers wages	
Total cost of materials	
Materials, parts, containers, packaging, etc. used	Materials, parts, containers, packaging, etc used
Cost of resales	Contract work and resales
Contract work	Contract work and resales
Cost of purchased fuels	Purchased fuels and electricity
Purchased electricity	Purchased fuels and electricity
Quantity of electricity purchased	
Quantity of generated electricity	
Quantity of electricity sold or transferred	
Total value of shipments	Shipments
Value of products shipments	
Total miscellaneous receipts	
Value of resales	
Contract receipts	
Other miscellaneous receipts	
Value of interplant transfers	
Value added	Value added
Total EOY inventories	
Finished goods inventories, EOY	Net Inventories Shipped
Work-in-process inventories, EOY	Net Inventories Shipped
Materials and supplies inventories, EOY	
Total BOY inventories	
Finished goods inventories, BOY	Net Inventories Shipped
Work-in-process inventories, BOY	Net Inventories Shipped
Materials and supplies inventories, BOY	
Total capital expenditures (new and used)	
Capital expenditures: buildings & other structures (new and used)	Capital expenditures: buildings and other structures (new and used)
Capital expenditures: machinery and equipment (new and used)	
Capital expenditures: autos, trucks, etc. for highway use	Capital expenditures: machinery and equipment (new and used)
Capital expenditures: computer and data processing equipment	Computer hardware, software, and other equipment
Capital expenditures: all other machinery and equipment	Capital expenditures: machinery and equipment (new and used)
Total depreciation	Depreciation
Total rental payments	
Buildings rentals	Capital expenditures: buildings and other structures (new and used): Rental
Machinery rentals	Capital expenditures: machinery and equipment (new and used): Rental
Total other expenses	
Temporary staff and leased employee expenses	Other costs
Expensed computer hardware and other equipment	Computer hardware, software, and other equipment
Expensed purchases of software	Computer hardware, software, and other equipment
Data processing and other purchased computer services	Professional, technical, and data services
Communication services	Communication services
Repair and maintenance services of buildings and/or machinery	Maintenance and repair
Refuse removal (including hazardous waste) services	Refuse removal
Advertising and promotional services	Other costs
Purchased professional and technical services	Professional, technical, and data services
Taxes and license fees	Other costs
All other expenses	Other costs

Volume of Production=total costs (blue plus orange plus red plus green plus gold)
Net Inventories Shipped=sum of EOY finished goods and work-in-process inventories less the sum of BOY finished goods and work-in-process inventories

Table A.2: Total Supply Chain Values for Industries Relevant to Additive Manufacturing, $million 2011

	Communication Services	Other Costs	Refuse Removal	Computer Hardware, Software, and other Equipment	Professional, Technical, and Data Services	Payroll, Benefits, and Employment	Employment
Motor vehicles	167	14 995	391	542	1 408	47 238	651
Aerospace	155	7 682	177	689	3 142	34 987	333
Industrial/business machines	528	22 773	475	1 351	2 863	70 427	965
Medical/dental	151	8 903	123	477	1 409	21 533	290
Government/military	52	1 813	57	192	341	11 024	82
Architectural	121	4 823	101	226	494	17 631	302
Consumer products/electronics, academic institutions, and other	1 854	64 203	1 629	4 540	7 819	205 764	2 894
Total	3 028	125 192	2 952	8 016	17 477	408 603	5 516

	Contract Work and Resales	Purchased Fuels and Electricity	Maintenance and Repair	Volume of Production	Net Inventories Shipped	Depreciation	Net Income	Shipments	Capital Expenditures: Buildings and Other Structures	Capital Expenditures: Machinery and Equipment Used	Capital Expenditures: Materials, Parts, Containers, Packageing, etc.	Value Added (ASM)	Additive Manufacturing's Share of Shipments
Motor vehicles	9 300	2 806	2 116	399 482	-1 108	9 695	37 220	445 289	2 345	10 686	307 489	126 751	0.021%
Aerospace	7 739	1 257	611	124 628	-5 920	2 181	36 812	157 701	1 323	2 515	65 064	90 216	0.036%
Industrial/business machines	20 494	2 704	2 218	305 856	-4 414	6 477	57 815	365 735	3 802	8 876	169 346	177 486	0.014%
Medical/dental	4 446	525	450	63 730	-17	1 987	23 819	89 519	1 230	1 850	22 633	61 932	0.079%
Government/military	3 386	221	117	29 132	-323	468	3 507	32 784	380	407	9 482	18 350	0.086%
Architectural	4 934	723	441	64 799	-408	1 458	6 338	72 187	1 090	1 469	32 747	34 162	0.019%
Consumer products/electronics, academic institutions, and other	48 869	9 592	6 319	726 410	-5 886	22 966	152 219	895 710	11 589	26 634	338 544	505 513	0.018%
Total	99 168	17 828	12 273	1 714 038	-18 076	45 234	317 730	2 058 926	21 760	52 437	945 304	1 014 411	0.023%

49

Table A.3: Supply Chain Values for Additive Manufacturing by Industry, $million 2011

	Communication Services	Other Costs	Refuse Removal	Computer Hardware, Software, and other Equipment	Professional, Technical, and Data Services	Payroll, Benefits, and Employment	Employment	Capital Expenditures: Buildings and Other Structures	Capital Expenditures: Machinery and Equipment	Materials, Parts, Containers, Packageing, etc. Used
Motor vehicles	0.02	1.6	0.04	0.1	0.2	5.1	70	0.3	1.2	33.1
Aerospace	0.03	1.5	0.03	0.1	0.6	6.6	63	0.2	0.5	12.3
Industrial/business machines	0.04	1.7	0.03	0.1	0.2	5.1	70	0.3	0.6	12.3
Medical/dental	0.06	3.7	0.05	0.2	0.6	8.9	120	0.5	0.8	9.4
Government/military	0.02	0.8	0.03	0.1	0.2	5.0	37	0.2	0.2	4.3
Architectural	0.01	0.5	0.01	0.0	0.1	1.8	31	0.1	0.2	3.3
Consumer products/electronics, academic institutions, and other	0.17	5.9	0.15	0.4	0.7	19.0	267	1.1	2.5	31.3
Total	0.4	15.7	0.3	1.0	2.5	51.5	658.3	2.6	5.8	106.0

	Contract Work and Resales	Purchased Fuels and Electricity	Maintenance and Repair	Volume of Production	Net Inventories Shipped	Depreciation	Net Income	Shipments	Value Added (ASM)	Additive Manufacturing's Share of total Shipments
Motor vehicles	1.0	0.3	0.2	43.1	-0.1	1.0	4.0	48.0	13.7	0.01%
Aerospace	1.5	0.2	0.1	23.5	-1.1	0.4	7.0	29.8	17.0	0.02%
Industrial/business machines	1.5	0.2	0.2	22.2	-0.3	0.5	4.2	26.6	12.9	0.01%
Medical/dental	1.8	0.2	0.2	26.5	0.0	0.8	9.9	37.2	25.7	0.04%
Government/military	1.5	0.1	0.1	13.1	-0.1	0.2	1.6	14.8	8.3	0.05%
Architectural	0.5	0.1	0.0	6.6	0.0	0.1	0.6	7.4	3.5	0.01%
Consumer products/electronics, academic institutions, and other	4.5	0.9	0.6	67.1	-0.5	2.1	14.1	82.7	46.7	0.01%
Total	12.3	2.0	1.4	202.1	-2.3	5.2	41.3	246.1	127.7	0.01%

Appendix B: Equations and Assumptions

The approximations for U.S. additive manufacturing activity rely on the assumption that the U.S. share of additive manufacturing systems sold equates to the share of products produced using additive manufacturing systems. This is represented as the following:

$$R_{US} = \frac{S_{US}}{S_G} R_G$$

Where:
R_{US} = Revenue for additive manufacturing activities in the U.S.
S_{US} = Cumulative number of additive manufacturing systems sold in the U.S. between 1988 and 2001
S_G = Cumulative number of additive manufacturing systems sold globally between 1988 and 2001
R_G = Revenue from the global sale of parts produced from additive manufacturing systems

Shipments of additive manufactured parts and products by category (see Table 4.1) was estimated by assuming that the percent of additive manufacturing that each category represents is the same for the U.S. as it is globally. The calculation is represented as the following:

$$R_{US,x} = \frac{R_{G,x}}{R_G} R_{US}$$

Where:
$R_{US,x}$ = U.S. revenue for additive manufacturing activities for category x
$R_{G,x}$ = Global revenue for additive manufacturing activities for category x
R_G = Global revenue for additive manufacturing
R_{US} = Revenue for additive manufacturing activities in the US

www.ingramcontent.com/pod-product-compliance
Lightning Source LLC
Chambersburg PA
CBHW081740170526
45167CB00009B/3885